FOREWORD

National programmes and policies concerning the civil use of plutonium are quite diverse. Some countries have seen plutonium as a valuable resource, while others consider that it should be retained in the spent fuel. The operation of large reprocessing plants has resulted in the separation of large quantities of plutonium which have, up to now, only been partly used. From the beginning of the interest in the use of nuclear energy, it was envisaged that plutonium would be separated and recycled into fast breeder reactors in order to make full use of the fission energy potential of uranium fuel. More recently the slower growth of the use of nuclear energy has led to reconsideration of the timing for the introduction of fast reactors. For a variety of reasons, concerns have been expressed both by government and other commentators about the existence and use of the accumulated plutonium stocks.

The NEA Committee for Technical and Economic Studies on Nuclear Development and the Fuel Cycle took the view that it would be timely and helpful to the public debate about these concerns to have a report on the experience to-date with the handling and use of plutonium in the civil nuclear industry, and also to review potential technologies that could be expected to become available in the future for the utilisation or disposition of plutonium. Accordingly, it set up a group of experts from countries that had made use of plutonium (including the Russian Federation), as well as from non-user countries, and from the IAEA and the European Commission. A full list of members of this *ad hoc* group is given in Annex 1. This report represents the consensus view of these experts.

In recent years, the United States and the Russian Federation have announced that they are to release plutonium which is excess to their national military needs.

This report does not specifically discuss the disposition of such material. However, because the technologies used in the civil sector have been developed to cope with a quite wide range of plutonium isotopic compositions, they could, for the most part, be adapted to handle material derived from surplus nuclear weapons.

This report has been prepared by the members of the expert group and is published under the responsibility of the Secretary-General of the OECD. It does not necessarily represent the views of participating countries or international organisations.

Acknowledgement

The help provided by Mr. H. Bairiot in carefully reviewing the final draft of the report is gratefully acknowledged.

EXECUTIVE SUMMARY

Plutonium is produced during the operation of uranium-fuelled reactors and is, as a result, a constituent of the spent fuel produced by them. The policies and programmes of both governments and utilities concerning the use of plutonium are quite diverse. Some involve separating it from spent nuclear fuel, others do not. During the past years a sizeable amount of plutonium has been separated but only used on a limited scale. Currently, its use is growing. There is clearly active public and government interest in ensuring that plutonium, whatever its origin, is managed safely with no appreciable risk to mankind and the environment.

In early 1994, an *ad hoc* expert group, with a membership drawn from 15 countries and three international organisations, was formed to examine, review, and to provide an account of technical options – currently available or under development – for managing separated plutonium.

The study of the expert group shows the extent to which several relevant technologies are already in use on an industrial scale. Consequently, in the short and medium terms, *i.e.* the period comprising the next 15 to 20 years, plutonium can be effectively recycled in thermal reactors. Successful implementation of mixed uranium-plutonium oxide (MOX) recycling programmes, which are under way in a number of countries, would result in an equilibrium between plutonium production and consumption in OECD countries and would eventually reduce civil plutonium stocks. However, surplus quantities will need to continue to be safely stored. The necessary technologies are commercially available, proven, safe, and can be safeguarded.

In the longer term, depending on the evolution of nuclear policies, a similar route to that indicated above could be pursued. In addition, plutonium may be used more efficiently in fast and other reactor types or may be conditioned to a form appropriate for final disposal. The latter techniques, presently under research and development, would need to be fully demonstrated and accepted prior to implementation.

It should be noted, however, that successful implementation of such technical policies would depend on a complex array of interrelated considerations of national and international policy issues which were beyond the scope of the expert group.

The management of separated plutonium presents no major technical difficulties, but is merely a matter of applying existing technology to the minimisation of any plutonium stocks.

TABLE OF CONTENTS

LIST OF FIGURES

LIST OF TABLES

Chapter 1

INTRODUCTION

1.1   Background

The increasing quantities of separated civil plutonium and the postponement or the abandonment of plans for Fast Breeder Reactors (FBRs) have resulted in a growing interest, in a number of OECD countries, in recycling plutonium in Light-Water Reactors (LWRs).   While natural uranium requirements, enrichment and fuel fabrication capacities and requirements, reprocessing capacities and spent fuel production can be reliably predicted on the basis of the present operating and scheduled nuclear power capacity until about the year 2005, future separated plutonium inventory forecasts have significant associated uncertainties and are quite sensitive to MOX fuel fabrication and spent fuel reprocessing assumptions.

It is recognised that civil plutonium management and use are not only related to technological, economic and safety considerations, but also have wider non-proliferation, political, national and international policy, and public acceptability implications (1).

Management of civil plutonium is seen as one of the many challenges the world nuclear community has to face today.

Technical Committees of the OECD Nuclear Energy Agency (NEA) have many years of accumulated experience in dealing with plutonium issues, including among others, economics, safety, related technological and scientific aspects and recycling logistics.   Relevant work is usually performed by *ad hoc* expert groups consisting of nominated high-level national experts and is primarily addressed to OECD member governments.

Past and present NEA plutonium work has been, essentially, targeted at economic aspects and the need to identify suitable technical solutions, despite the existing large political uncertainties associated with their implementation.   Institutional aspects, non-proliferation and physical security are treated in the International Atomic Energy Agency (IAEA) and in other fora.

In the mid-1980s the NEA's Committee for Technical and Economic Studies on Nuclear Development and the Fuel Cycle (NDC) conducted a study on plutonium and published in 1989 the resulting report: *"Plutonium Fuel - an Assessment"* (1).   This provided the facts and then current views about plutonium and its civil use in the short and medium terms.   It addressed questions influencing the choice of fuel options and illustrated how economic and logistic assessments of the alternatives could be undertaken.

Plutonium handling and safety issues, based on available experience in OECD countries, were addressed in the 1993 NEA study on the *"Safety of the Nuclear Fuel Cycle"* (2).   The report underlined the extreme care which has been taken in both the design of plutonium handling facilities and the detailed operational procedures that have been adopted, and noted their very good safety records.

Regarding the economics of plutonium use, the 1994 NEA study on the *"Economics of the Nuclear Fuel Cycle"* (3) included updated analyses of the economics of MOX fabrication and of the entire reprocessing cycle.

The recently published NEA study on the *"Physics of Plutonium Recycling"* (4) reviewed the physics issues associated with plutonium recycling, and specifically addressed the impact of amounts, compositions and toxicities of plutonium, transuranic flows and inventories arising from the various options chosen for the back-end of the fuel cycle.

1.2 Scope and objectives of this study

National programmes and policies concerning the civil use of plutonium are quite diverse. Public opinion is quite sensitive in almost all countries irrespective of the material's original nature, be it civil or military. Although the principal technologies are currently available to reduce the quantities of separated civil plutonium, implementation of the national strategies which would lead to such reductions is linked to public perception (5).

The growing plutonium inventories may well exacerbate concerns about the continued use of nuclear energy expressed in certain quarters. To the extent that it is seen as desirable that the option of using plutonium as fuel for nuclear reactors should be maintained in the future, it should be demonstrated that civil stocks of plutonium will continue to be managed well in the medium term and the following topics should be addressed:

- the availability of and experience in using technology for all segments of the plutonium recycle route; and

- the potential for further improvements in technology, aiming, for example, at even higher standards of environmental protection, safety, worker health and reduced costs.

With those considerations in mind, an *ad hoc* expert group was assembled, in early 1994, under the auspices of the NEA's Nuclear Development Committee, with the task of identifying, examining and evaluating the broad technical questions related to plutonium management. Recognising that this was a subject of interest to all countries, whether or not they had in stock separated plutonium or any intention of using plutonium, the following countries nominated participants to the expert group: Australia, Belgium, Canada, France, Germany, Ireland, Italy, Japan, Korea, the Netherlands, Norway, Switzerland, the United Kingdom and the United States. Exceptionally, experts from Russia were invited to participate because of the accumulated experience in that country with plutonium production, handling and use. The IAEA and the European Commission were also represented. The list of group members is given in Annex 1. The expert group was chaired by Mr. C.J. Joseph.

The objective of the expert group originated from some basic facts: *"plutonium arises in operating reactors; most of it is held in spent fuel; a growing amount has been separated but not yet used; there is public interest in the management of this plutonium, including any that may become available to the civil market from ex-military uses"*.

Regarding the terms of reference of the expert group, it was agreed that its study would be *"concerned with the technical options for management of this plutonium, focusing on the following topics:*

a) *Technologies have already been implemented which provide for medium-term storage of plutonium or for recycling the plutonium through reactors. A review will be provided of experience gained with them and a technical commentary on their potential deployment over the next 20 years or so.*

b) *In the longer term, these technologies may be joined by a further range which are, in some cases, already under development. A technical review of the additional options that may become available will be provided."*

As mentioned before, neither institutional matters nor military plutonium as such were considered by the expert group, but note was taken of specific problems or opportunities associated with plutonium of different isotopic compositions. The expert group did not perform a detailed analysis of plutonium arisings, nor did it examine the economics of the various technological options considered.

1.3   Spent fuel management

The management of reactor spent fuel, in what is commonly known as the "back-end" of the fuel cycle, has always been one of the cycle's most important stages. Decisions about this stage are now among the most challenging facing countries that operate nuclear reactors (6).

In 1995, the annual world-wide spent fuel arisings from all types of power reactors amounted to about 10 500 t HM, giving an estimated cumulative total of over 165 000 t HM. About 110 000 t HM are presently being stored. The quantity of accumulated spent fuel is 20 times the present total annual reprocessing capacity (7). The projected cumulative amount of spent fuel generated by 2010 will reach 300 000 t HM. About 100 000 t HM are scheduled or planned to be reprocessed, leaving about 200 000 t HM of spent fuel to be stored by 2010. Since the first large-scale geological repositories for final disposal of spent fuel are not expected to be in operation before the year 2010, the indications are that interim storage will be the primary option for the next 20 years (7).

As far as plutonium utilisation is concerned, there are, at present, two broad and quite diverse policies world-wide: while a number of countries (e.g. Canada, Sweden, United States) tend to consider that plutonium is best retained in the spent fuel in long-term storage followed by final disposal in deep geological formations, others (e.g. Belgium, France, Germany, Japan, Russia, Switzerland, United Kingdom) have seen it as a valuable resource. The latter group of countries are pursuing the reprocessing option and are operating or constructing reprocessing plants or have contracts for reprocessing their spent fuel abroad. The reasons for selecting a national strategy are specific to each country and are based primarily on economic and political considerations (8). Countries with monitored retrievable storage and direct disposal of spent fuel as a back-end strategy have selected this option for non-proliferation and/or economic reasons, and then for social and cultural considerations (8). Countries with recycling and reprocessing as a back-end strategy typically have limited energy resources and a strong nuclear programme (8). Many countries, generally having smaller nuclear power programmes, are currently deferring a decision on which of the above-mentioned strategies to select and are storing their spent fuel. Annex 3 contains a table from the 1996 NEA *"Brown Book"* which summarises the nuclear power programmes and their evolution in OECD countries.

It is worth noting that some countries employ different spent fuel management approaches for different fuel types. In others, a single spent fuel management approach is currently being followed, but future options in which a variety of approaches would be applied have been adopted or are currently being evaluated. For example, many utilities in Belgium and Germany currently reprocess part of their reactor spent fuel, but may in future choose to dispose their spent fuel without prior reprocessing.

Selection of national strategies for the back-end of the fuel cycle involves an integrated consideration of economics, technical issues, high-level waste management aspects, future societal needs, continuity of existing programmes, safety and environmental factors, concerns about acceptance, non-proliferation issues and other socio-political considerations (1, 8).

An international approach on spent fuel management and plutonium use should fully reflect the diverse national political, economic and technical situations, as well as geographical particularities and global security issues.

1.4   Reprocessing

Spent fuel reprocessing is a proven technology which is offered commercially on an international basis by COGEMA (France) and BNFL (United Kingdom). Japan is actively developing plans to build a commercial reprocessing plant. In Russia, a multi-purpose reprocessing plant has been in operation for nearly 20 years and a commercial reprocessing plant has been proposed.

Reprocessing involves dissolving the spent fuel to enable the re-usable plutonium and uranium content to be separated from the residual waste fission products and minor actinides. Spent fuel from a large modern PWR, discharged at an average burn-up of 42.5 GWd/t, typically contains 1.15 per cent (by weight) plutonium, 94.3 per cent uranium and 4.55 per cent waste products (3). The separated uranium may then be re-enriched prior to re-use and the separated plutonium may be incorporated into MOX fuel. In this manner, fuel materials containing about 30 per cent of the potential energy of the initial fuel can be re-utilised in thermal reactors, and more if fast reactors are used.

In 1995, the total world annual reprocessing capacity for all fuel types amounted to 5 565 t HM. All operating reprocessing plants use Purex technology. No particular difficulties are anticipated in dealing with uranium or MOX fuels of high burn-up (up to 60 MWd/kg HM). The future total reprocessing capacity and throughput will depend on the future demand for reprocessing services. The projected reprocessing capacity would remain relatively constant over the period 1995 to 2010 (7). The proven Purex technology is unlikely to be replaced, although some modifications and improvements may be introduced. The main areas of development in reprocessing are reduction of capital and operating costs, reduction of waste volumes and increased automation.

1.5   Plutonium arisings and consumption

Plutonium is generated (and is partly burned) during the operation of uranium-fuelled nuclear reactors and forms part of their spent fuel. Currently, some 50 t of plutonium are generated worldwide every year in spent fuel. By the end of 1994, about 700 t of plutonium had been formed in commercial nuclear fuel; by the year 2000 about 1 010 t and by the year 2010 about 1 520 t of plutonium will have been generated (7).

Some of the plutonium contained in reactor spent fuel has been separated and a part of it (approximately one third) has, up to now, been used to manufacture MOX fuel for LWRs and experimental and prototype FBRs, but the major part of the separated plutonium is currently stored, mainly at the British, French and Russian reprocessing sites.

Plutonium arisings are determined by the amount of spent fuel which is contracted for reprocessing. The two main industrial parameters regarding the consumption of separated plutonium are the MOX fuel fabrication capacity and the number of reactors which are licensed to use MOX fuels. These factors explain how demand and supply can be balanced in the coming years.

Currently, MOX fuel fabrication capacities in OECD countries represent a flow of 190 t HM per year. This level corresponds to some 10 to 12 t of plutonium used in MOX per year. The available MOX fabrication capacity would reach approximately 400 t HM in the year 2000, corresponding to some 25 to 30 t of plutonium per year (with the planned increase of the plutonium percentage in the fuel).

The number of light-water reactors licensed at present to load MOX fuel is 32. Utilities involved in fuel reprocessing and recycling are currently undertaking the necessary steps to increase the number of their plants, which are licensed to use MOX fuel, to a total of 50 to 60 by the year 2000. Thus, the future MOX plant fabrication capacities would allow a regular flow of plutonium to such plants and result in an equilibrium of plutonium arisings and consumption around the year 2000.

In 1995, 22 t of plutonium were separated and 8 t of plutonium were used in LWRs and in breeder reactor development programmes. The imbalance between the separation and the use of plutonium had resulted in an inventory of separated civil plutonium of about 126 t at the end of 1995 (7).

According to an IAEA analysis (7), the rate of separation of civil plutonium would continue to exceed its rate of use up to the year 2001. By the end of the year 2000, the estimated inventory would have reached about 186 t plutonium. Beyond 2001, the inventory is expected to decrease at a rate of about 10 t per year over the subsequent ten years. Towards the end of the next decade, the rate of decrease of the inventories of separated plutonium will decline, driven by both the increasing working inventories of plutonium needed as MOX and fast reactor fuel production rates rise, and the increasing quantities of plutonium in both the UK and Russia. The results of the analysis are quite sensitive to the start-up dates and throughputs of MOX facilities, as well as to rates of plutonium use in fast reactors. The IAEA analysis noted that, under existing contracts, the utilisation of the full capacity of reprocessing plants is not assured beyond 2000; and in the event that MOX facilities are not ready on time, utilities may delay reprocessing to avoid the costs of storing plutonium. Such delay may also avoid the costs of americium removal from aged plutonium (7).

The analysis indicated that there are significant uncertainties in the prediction of the maximum amount of plutonium, the time of occurrence of that maximum and its subsequent rate of decline (7).

The situation regarding plutonium stocks, and plans which are underway to use them, differ from country to country. In some countries MOX programmes are already actively implemented, in others, though, recycling of separated plutonium is not expected to take place in the short term. More information about country programmes can be found in Chapter 4.

Under the first and second Strategic Arms Reduction Treaties (START I and II) and unilateral pledges made by American and Russian Presidents, many thousands of US and Russian nuclear weapons are now planned to be retired within the next decade. As a result, 50 or more tonnes of weapons-grade plutonium on each side are expected[1] to become surplus to military needs (9). It is not unlikely that part or all of this surplus plutonium may eventually reach the commercial market. Civil fuel cycle infrastructures, services and programmes may, therefore, need to be developed and adjusted accordingly to cope with this material.

1.6   Other factors affecting the management of plutonium

Economic considerations

The economics and the related logistics of plutonium recycling have been well studied over the years (1, 3). It is understood that other plutonium management methods which are currently under investigation have large associated economic uncertainties.

Safety and physical security

The risks and hazards associated with separated plutonium handling, storage, transport and use are well understood and technologies are available to conduct these activities safely.

Civil plutonium handling facilities have very good operational and safety records in OECD countries (2) and provide high standards of protection for the environment and public health. Plutonium confinement and storage, and plutonium transportation either in solution or solid forms follow well established practices and international guidelines (12); further improvements are expected in the near future.

Various national security arrangements (use of barriers, detection measures, guards, etc.) are aimed at physically protecting isolated plutonium and preventing its diversion for non-peaceful uses (IAEA INFCIRC/225, Rev. 2 guidelines).

Non-proliferation

A number of international, regional and bilateral agreements exist in order to ensure that civil nuclear materials and equipment, such as plutonium and plutonium handling facilities, are subject to safeguards to prevent their illicit use for the manufacture of nuclear weapons or explosive devices. Most countries with civil nuclear power programmes also have their own safeguards system. Civil plutonium separation, transport, storage and use are, therefore, under strict government control and international surveillance.

---

1   The United States has placed more than 10 tonnes of excess fissile material under international safeguards (10) and has also declared 200 tonnes of fissile material, of which 38.2 metric tonnes of weapons-grade plutonium, surplus to national security needs (10, 11).

Reprocessing and MOX fuel fabrication plants, as well as civil plutonium storage facilities, can be satisfactorily safeguarded. Although plutonium handling facilities exist in several countries, large commercial reprocessing and MOX fuel fabrication plants are located within the territory of the European Union; the IAEA, as well as EURATOM have gained substantial experience over the years in safeguarding these parts of the fuel cycle (13). Procedures used are comprehensive and well established, and even more efficient methods are currently being pursued.

Although the idea of management of excess plutonium under an international scheme has been given active consideration in the past, this option has not been formally pursued owing to various political and national policy reasons. At present, relevant ideas continue to be explored and evaluated.

Related general policy issues

It is increasingly recognised that political and associated proliferation concerns in dealing with plutonium are much more complex than related technical questions. Of importance are also the evolution of different energy supply options and concerns about the environment that need to be addressed by governments. Quite often decisions affecting the nuclear fuel cycle need to be balanced against other national and international policy concerns (14).

Transparency of civil plutonium use and non-proliferation controls are essential in obtaining public confidence and support for national policies. It is noted that significant efforts have been made, during the past three years, by a number of countries to enhance transparency regarding both their civil plutonium use and stocks.

Noting that since the public does not, in general, see a real difference between military and civil plutonium, national and international activities aimed to account, secure and manage ex-military (and, of course, civil) plutonium safely should be clearly presented. Although management and eventual disposition of surplus military plutonium will most likely take place within a complex international context, this issue is currently viewed primarily as a bilateral question to be addressed by Russia and the United States.

1.7   Report organisation

Chapter 2 of the report deals with those fuel cycle steps involving plutonium handling and use, namely: finishing, packaging, storage, transportation, MOX fabrication, plutonium use in MOX fuel in thermal and fast reactors, MOX fuel reprocessing and plutonium purification. Experience gained and near-term evolution are reviewed. Chapter 3 addresses future developments regarding plutonium burning in current and advanced reactor concepts, as well as plutonium immobilisation and geological disposal. Country programmes are presented in Chapter 4.

Chapter 2

EXPERIENCE GAINED AND NEAR-TERM EVOLUTION

2.1   Introduction

Production of civil plutonium is now into its fourth decade and equipment and systems are at an advanced state in terms of:

- quality control;
- minimisation of effluents;
- minimisation of doses to operators and the public;
- radiological and non-radiological safety;
- safeguards; and
- physical protection.

Safety considerations are of prime importance in the design, construction and operation of plutonium facilities. All such undertakings are in strict compliance with the conditions of the Nuclear Site Licence issued by the Competent Authority, e.g. the Nuclear Installations Inspectorate in the United Kingdom and the Direction de la Sûreté des Installations Nucléaires in France. Enforcement of the Site Licence, and compliance with documentation issued by the company to support safe operations, is carried out by the Competent Authority who operate routinely on all nuclear licensed sites.

In modern reprocessing plants like UP2-800, UP3 and THORP, the inventory at all stages of the process is routinely determined from central computers linked to an array of direct and/or indirect plutonium monitors installed into the plant. Some of the monitors are the property of EURATOM and the IAEA who verify the data supplied by the operators. EURATOM is also involved in the design and construction phases of such plants.

Because of the radiotoxic nature of plutonium, the facilities concerned with its handling operate in alpha tight environment with automated and remotely controlled equipment.

2.2   Finish, packaging and storage

Finish

Plutonium separation at the major reprocessing facilities in the UK and France is based on the precipitation of plutonium oxalate from plutonium nitrate solution after chemical separation from fission products and uranium.

After blending of the nitrate solution to ensure a homogeneous batch with respect to concentration, isotopics and impurities, the solution is conditioned to its +4 oxidation state to minimise the solubility of the oxalate and hence losses (Figure 1). A slight excess of oxalic acid solution is then blended with the nitrate at a controlled temperature and this causes immediate precipitation of hydrated plutonium oxalate. The crystals are allowed to "age" for at least ten minutes to allow growth. The precipitate is then filtered and washed.

Figure 1.   Outline flowsheet for Pu(IV) oxalate precipitation

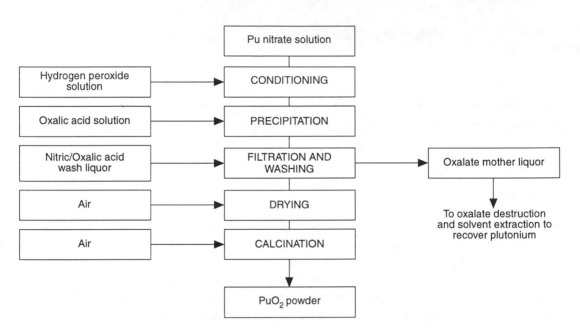

The solution passing through the filter contains residual plutonium. This solution is concentrated by evaporation and then refluxed with concentrated nitric acid to destroy residual oxalic acid before the plutonium stream is fed back into the chemical separation area.

The precipitate is continuously scraped from the filter from where it drops into a drying furnace where it is moved along by an internal Archimedes screw. Excess water is removed, followed by water of crystallisation, followed by thermal decomposition of the oxalate to give plutonium dioxide. The plutonium dioxide then passes to a second "calcination" furnace where the temperature is controlled at around $600^{\circ}C$ to reduce the surface area of the powder to around 10 m^2 per gram. A high specific surface area improves the homogeneity of MOX fuel when fabricated, but too high a surface area increases the capacity for adsorption of moisture.

In Japan, plutonium is co-converted with uranium in a microwave assisted thermal denitration process in the current small-scale finishing and fabrication plant at Tokai-Mura and planned commercial-scale reprocessing plant at Rokkasho-Mura. Plutonium nitrate solution from reprocessing is blended with uranyl nitrate solution and this is transferred to crucibles in which it is thermally decomposed using microwave energy. The resulting master blend of uranium and plutonium oxide powder is removed from the boats and milled and heat conditioned, then blended down to the required fissile content for subsequent fuel fabrication.

Packaging

Packaging is required to provide adequate containment of the plutonium dioxide powder, to avoid the possibility of criticality, and to assist in removal of decay heat.

Plutonium dioxide is cooled and packaged under an inert atmosphere to minimise the amount of water which is adsorbed onto the surface area of the powder.

The radiation from plutonium dioxide depends on its isotopic composition and the impurities in it and increases with the burn-up of the original fuel. For plutonium from low burn-up fuel, e.g. from the UK's Magnox reactors, the plutonium is weighed into three-litre aluminium screw top bulpitt containers which are then posted out of the alpha containment glovebox in polythene which is heat sealed. The polythene intermediate packages are placed inside deep drawn stainless steel cans and lids are sealed on by resistance welding.

In the COGEMA's plant located in Marcoule, plutonium obtained from the reprocessing of gas-cooled reactor spent fuel is introduced into stainless steel cans. Each can is filled with a maximum of 3.4 kg plutonium dioxide. After filling with plutonium, the cans are sealed by crimping, and they are weighed. The packaging is then completed by introducing the cans into two successive stainless steel containers.

While plutonium derived from Magnox fuel contains, typically, 0.2 per cent ^{238}Pu, plutonium from high burn-up oxide fuel will contain up to 2 per cent of the ^{238}Pu isotope, which is a strong heat generator. Consequently, plastic intermediate packages cannot be used.

For plutonium dioxide from THORP, the aluminium inner can is replaced by a three litre stainless steel screw top can. This inner can is inserted in a close-fitting stainless steel intermediate can in a sphincter seal to the alpha containment. A bung is placed in behind the inner can and a laser weld is made by rotating the can. The welded intermediate can is then placed inside a stainless steel outer can and a lid sealed by resistance welding. The bung in the sphincter seal is displaced by the next intermediate can. The three layers of close fitting stainless steel improve the thermal conductivity of the package, as well as strength, but, with up to 20 W per kg of heat generated, the centre line temperature of the oxide can still reach 500°C.

Similarly, the basic principles of plutonium packaging in France are:

- to place the PuO_2 powder in a stainless steel can which is sealed by crimping; and
- to achieve a complete containment through two successive stainless steel containers with welded lids.

Packaging and transport equipment is constantly improving, bearing in mind the objective of conditioning larger quantities of plutonium and continuing to meet safety and security standards. The inner can holds some three kg of heavy metal. Five of these inner cans are loaded into a larger stainless steel outer can. This outer can is over-contained in an outer storage container which acts as secondary containment.

Proof testing of package designs involves destructive testing by pressurisation of empty containers under normal and faulty conditions at normal and at elevated temperatures. Drop testing of packages followed by pressure testing is also undertaken.

The final product is leak tested to demonstrate compliance with international transport requirements (less than 10^{-6} atmos/cm^{3}s).

Storage

Packages of plutonium dioxide are stored within massive concrete cells or pits to afford protection against external hazards such as seismic events and aircraft crashes. The stores are also designed to avoid criticality, and to take account of the release of heat and the physical containment and protection of the plutonium.

At Sellafield, cans of plutonium dioxide are stored in re-entrant tubes accessible only through loading ports fitted with shield plugs in the operating face of the cell. Loading of containers to the channels is by means of a shield block and remote mechanical system. Cooling of the store's contents is necessary due to the radioactive decay heat generated and this is undertaken by forced cooling, although the tall racks generate their own chimney effect. During design and commissioning, the safety of the store during any loss of ventilation was demonstrated by utilising electrical heaters in the can positions.

Inspection of the plutonium packages is carried out remotely using a mobile arrangement of television monitors and thermocouples.

At La Hague the following safety aspects have been taken into account in the design and operation of the BSI ("Bâtiment de Stockage Intermédiaire", *i.e.* intermediate storage building):

- criticality safe plutonium container network spacing and requirements for keeping a permanent store geometry;
- the containers have been designed in such a way that they can withstand a loss of ventilation in the facility;
- in addition to the quality of the containers there have been improvements in the reliability of the handling equipment (risk of dropped loads); and
- the store is seismically qualified.

Experience with storing low burn-up Magnox PuO$_2$ for long periods (up to 30 years), mainly in the United Kingdom (15), has shown that periodic surveillance and recanning is necessary, due to the radiation induced deterioration of the polythene intermediate bags. As outlined before, modern packaging comprises only successive stainless steel cans with welded lids. They prevent deterioration of the containment and no storage time limit would arise from radiation damage, even for very long-term storage (e.g. 100 years or more (16)). However, because of inadequate packaging processes which have been used, the integrity of the containment can be compromised (17). This has been in particular assessed for high ^{239}Pu type of plutonium (18). During prolonged storage plutonium ages due to the decay of ^{241}Pu into ^{241}Am. After 100 years storage, 99.2 per cent of the ^{241}Pu would have disappeared and, because ^{241}Am has a long half-life (458 years), the radioactivity of the stored plutonium would be high, especially if its ^{241}Pu content was initially high. Such plutonium would have further lost much of its fuel value. If recovery of such plutonium was required, either for utilisation as MOX or for final disposal, purification and/or handling would be complicated.

For plutonium containing high proportions of strong alpha emitters (in decreasing order of importance: ^{238}Pu, ^{241}Am, ^{240}Pu, ^{239}Pu, ^{241}Pu), the helium pressure build-up in the inner can may become a safety limiting feature. For very long-term storage, the can wall thickness and/or internal void volume must, therefore, be engineered, taking into account the isotopic composition of the plutonium to be stored. In cases of high helium pressures at the end of the storage period, the opening and even handling of the pressurised packages would present a radioactivity release hazard, requiring elaborated equipment and facilities to handle the packages. In this respect, the isotopic composition of the plutonium is the main factor limiting the time period over which long-term storage can be conducted.

2.3   Transportation

Introduction

Civil plutonium has been separated in reprocessing plants at Sellafield and Dounreay in the United Kingdom, Cap La Hague in France, Tokai-Mura in Japan and RT1 in Russia. Plutonium, in any of its forms, has been safely transported, for more than 35 years, internally within Europe, Russia and Japan, and internationally from the UK and France to Japan. Transport by road, rail, air and sea have all been used under international regulations inside approved transport containers.

Hazards associated with transportation of plutonium

The radiotoxicity of plutonium is well documented. Except for neutron irradiation, essentially due to ^{238}Pu, the main hazard of plutonium results from alpha radiation. The most hazardous physico-chemical form is fine particulated plutonium which enters the body by inhalation. Hence, the primary concern when transporting plutonium is to maintain the integrity of the containment.

Civil plutonium is usually in the form of nitrate when first separated and oxide for storage and use. Of these forms, oxide powder can become airborne and enter the body, chiefly through the lungs, although any reaching the alimentary canal will nearly all be excreted as it is not soluble. Nitrate solutions can become dry and respirable, and enter the body in solution through the gut.

Plutonium, above a critical mass, is capable of sustaining a nuclear chain reaction and so the quantity and its geometry must be controlled when it is in transit.

Because of its radiotoxicity and its potential for use in nuclear weapons, plutonium must be physically guarded against diversion from civil to military or terrorist purposes.

Dose rates associated with transportation of plutonium

Table 1 compares typical doses from the transport of different forms of radioactive material. It can be seen that the levels arising from the transport of plutonium are below those from other types of radioactive material.

Table 1 **Typical doses* from the transport of radioactive material**

		PuO$_2$ powder	MOX assembly	Irradiated fuel
Outer surface of package	gamma	10	26	140
	neutron	20	185	75
Outer surface of conveyance	gamma		9	25
	neutron		84	14
	total	1.7		
At 2 m from the conveyance	gamma		2	12
	neutron		15	7
	total	1.7		
In the driver's cabin	gamma		<1	4
	neutron		<10	1
	total	0.5		

*   *Notes*:
1.   Doses are in µSv/h.
2.   Typical shipment of PuO$_2$ in FS47 packaging (10 per vehicle).
3.   Measurements for typical shipment of MOX fuel assemblies from Belgonucléaire to NOK power plant in FS69/TNB176 package.
4.   Results of dose rate measurements of typical shipment of irradiated fuel from Doel power station (NTL3, 7 assemblies).
5.   Source: *"The impact of the 1991 ICRP recommendations on the transport of fissile materials and radioactive waste"*, I. Baekelandt and I. Lafontaine, May 1991.

Regulations

The transport of fissile materials and radioactive waste is carried out in accordance with the IAEA regulations for the safe transport of radioactive material (IAEA Safety Series No 6, *"Regulations for the Safe Transport of Radioactive Material"*, 1985), which have been incorporated into national laws. These regulations were first issued in 1961 and have been updated at regular intervals to take into account the progress in radiological protection and the evolution in transport practices. The next comprehensive revision of the transport regulations has been published in 1996.

The purpose of these regulations is to establish standards of safety which provide an acceptable level of control of the radiation hazards that are associated with the transport of radioactive materials. The radiation exposure of transport workers and of the general public is subject to the requirements specified in the IAEA Basic Standards for Radiation Protection (IAEA Safety Series No 9, *"Basic Safety Standards for Radiation Protection"*, 1982), which, in turn, are based on the ICRP Recommendations (ICRP Publication 60, *"Recommendations of the ICRP"*, annals of the ICRP 1(3), December 1994).

The aim of the IAEA regulations is to control:

- The external radiation hazard, by setting limits for dose rate at the external surface of packages and for the maximum radioactive content of these packages.

- The confinement of radioactive substances by setting limits for the maximum radioactive content of types of packages as a function of their mechanical integrity.

- The dissipation of heat by setting limits for the temperatures at the external surface of the package.

- Criticality hazards by setting limits for the quantity of fissile materials above which a criticality assessment must be performed during the design of the package.

Two international standards are laid down for the guidance of those involved in the storage and transport of radioactive materials. The first, *"The Physical Protection of Nuclear Material"* (INFCIRC/225/Rev. 3), recommends practical measures for physical protection of nuclear material in use, transit and storage. The second document, *"Convention of Physical Protection of Nuclear Material"* (INFCIRC/274/Rev. 1), provides for member states to give a commitment to take measures to protect material in transit and to co-operate and assist other member states in case of theft or diversion of nuclear material.

Whilst the technical details of how these standards are implemented on a national basis may vary from state to state, they all embody certain basic principles for the management of transportation and provide a comprehensive, international standard for the safe, secure transport of this class of materials which only approved transport organisations are permitted to perform.

Amongst the requirements are provisions to minimise the number of shipments, to reduce shipment and trans-shipment times and to avoid regular schedules and restrict prior knowledge of transport movements. Implicit in this is the requirement that the competence and reliability of the individuals and organisations involved must satisfy national government requirements.

The international standards provide tables of radioactive materials classified according to a physical security category. Each security category defines requirements to be taken into consideration for the protection of the material in transit.

INFCIRC/225/Rev. 3 lays down three security categories to which plutonium and uranium in their many forms may be allocated – from the highest Category I through to Category III. All plutonium materials from powder to fuel assemblies fall into Category I in quantities of two kg or more (unless exceeding 80 per cent ^{238}Pu).

Impact of changes to regulations

The 1996 changes to the IAEA regulations are:

- The introduction of type 'C' regulations, to increase the impact and fire test requirements of the package for air transport.

- Very Low Dispersability Material (VLDM). Material of a high A2 value, but by virtue of its form, *i.e.* MOX pellets would not require transport in type C packages due to it being inherently safer than powders.

- Increasing the Neutron Quality Factor (NQF), *i.e.* the number by which the value of the neutron radiation reading is multiplied to take into account its higher energy compared to gamma radiation.

Doubling the NQF to 20 will halve the current worth of shielding against neutrons – which is of particular importance for transporting plutonium, where two thirds of the radiation is from neutrons. The calculated effective dose rate is set in ranges under a Transport Index (TI), with higher TI values requiring more onerous transport arrangements. Thus, some shipments will move to a TI of over ten and will require carriage under "exclusive use", which will increase the number of shipments required. The majority of shipments of plutonium are currently carried under exclusive use due to the TI already exceeding ten and also for security requirements.

The enhanced drop and fire tests required by the proposed type C regulations currently under consideration by the IAEA are to be applied to the air transport of industrial quantities of plutonium, either as powder or fuel. Only packages capable of compliance, redesign or containing VLDM could be carried by air. Since at the present time these proposals are still under consideration, it would be inappropriate for member states to embark on any testing or redesign work until the final requirements are defined and the pass/fail criteria set by the regulatory process.

Types of packages

Type A packages are designed to withstand normal conditions of transport.

The content is limited to a value A2 if the material is in a dispersible form. This limit is relaxed to A1 if the material is in a non-dispersible form.

After tests to simulate minor mishaps (type A tests), there must not be a loss or dispersal of radioactive content nor loss of shielding integrity which would result in more than a 20 per cent increase in the radiation level at any external surface of the package.

Examples of A1 and A2 values are given in Table 2.

Table 2 **Examples of doses resulting from type A packages**

Radionuclide	A1 (GBq)	A2 (GBq)
^{238}Pu	2 000	0.2
^{239}Pu	2 000	0.2
^{240}Pu	2 000	0.2
^{241}Pu	40 000	10.0
^{241}Am	2 000	0.2

Type B packages are designed to carry in excess of the A2 value and withstand the consequences of severe accidents (type B tests).

Testing of transport containers

The rigorous tests for type B packages consist of:

- a drop from nine metres in the most vulnerable attitude onto an unyielding target, followed by;

- a drop from one metre onto a vertical punch in a position expected to cause the greatest damage, followed by;

- a fully engulfing fire with an average effective flame temperature of at least 800°C for 30 minutes, followed by;

- immersion under 15 meters of water for eight hours (200 m for irradiated fuel).

Performance of transport containers

Packages developed by BNFL (BNFL 1680) and Cogema (FS47) permit the shipment of high quantities of plutonium dioxide in powder form and are representative of the type of packages used by all agencies to transport plutonium.

The BNFL 1680 was designed to transport large quantities, around 50 kg, of plutonium dioxide arising from the Thermal Oxide Reprocessing Plant (THORP). It weighs approximately 2.5 t, is a little over a metre diameter and a metre long, and includes containment tubes made from very high strength stainless steel, ferralium. It can carry eight of the THORP product cans and dissipate over a kilowatt of heat and remain within the temperature limits of the IAEA regulations. The package is designed to minimise dose levels, by virtue of the timber shielding built in (a minimum of 200 mm) and by its capability of being loaded and unloaded remotely either vertically or horizontally. Each of the four carrier tubes is sealed by a cap with an outer viton 'O' ring and an inner metallic ring, providing a verifiable containment boundary.

The Cogema FS47 has a payload of 17 kg of plutonium dioxide and an overall weight of 1.5 t. Ten FS47 are put in a caisson equipped internally with a rack designed against chocking, which ensures lasting and facilitates handling operations. The FS47 contains successively:

- a cylindrical stainless steel container closed by a cover equipped with a double viton seal allowing testing of tightness before shipping;

- a neutron shield;

- a fire-proof protection; and

- an outer steel shell.

A shock absorber covers the entire top of the container.

Transport by road

Plutonium attracts the most onerous security requirements during transport. In Europe all plutonium materials from powder to fuel assemblies in Category I quantities (two kg or more) are transported in more or less the same way, since the regulations do not currently recognise the difference in form in which plutonium materials exist (for example MOX could be considered as possessing an intrinsic security characteristic resulting from the dilution of plutonium by up to 20 times in a uranium matrix).

The use of sea and air transport will usually involve road movements either at one or both ends since most facilities do not possess a rail head or sea terminal or airport within the site boundary. BNFL and Cogema have transported plutonium between nuclear installations by road using specially constructed vehicles incorporating the highest security standards. The vehicles are accompanied by armed escorts whose purpose is to further enhance the security of the shipment in addition to providing an extra communications channel.

Transport by rail

Rail transport of Category I materials outside nuclear sites is technically feasible under the guidelines provided in INFCIRC/225, and is covered by legislation in European nations.

Transport by air

Transport between countries in mainland Europe can generally be accomplished by road alone, but for particularly long shipments, and for current shipments to and from the United Kingdom (including some UK domestic shipments), air transport can be efficiently employed. By this means, the time needed to accomplish an international transport is reduced to a few hours, during which time the material is removed entirely from the public view with consequent security benefits.

The accident rate for civil aircraft is low; for severe accidents (defined as including at least one fatality and destruction of the aircraft) this rate is about 10^{-6} per flight. International research has shown that not all severe crashes will result in greater damage to the packages than that inflicted by the IAEA type B tests. Estimates range from 1 to 10 per cent of accidents. However, most packages greatly exceed the requirements of the IAEA tests, as shown by the results described earlier, where even at very high impact speeds IAEA leak-tightness is maintained and no real estimates of leakage of material over and above the IAEA requirement have been possible.

Several tonnes of mixed oxide fuel assemblies for the Dounreay Prototype Fast Reactor and residues have been transported by air since 1978, utilising the 1356 assembly container and the 2816C/E containers. A total of over four tonnes of plutonium, representing over 80 deliveries, have been made to overseas destinations from BNFL in the last 16 years. The material has been in the form of plutonium dioxide powder for at least 90 per cent of these transports, with the remainder being fabricated plutonium fuel in various forms.

Transport by sea

The UK regularly arranges movements of plutonium nitrate from Dounreay to Sellafield in a specially constructed vessel.

The US Department of Energy (DOE) has been shipping spent fuel from research reactors from foreign countries to the US since 1978, and over 300 of these shipments have taken place without incident.

About 1.1 t of fissile plutonium was transported from France to Japan to be used to fabricate fuel for the prototype fast reactor Monju. The transfer took place in a single non-stop shipment, taking from 11 November 1992 to 5 January 1993. The material was sealed into 133 French-made type FS47 multiple transport containers licensed by both French and Japanese Governments. The ship, being approximately 5 000 t and 100 m in length, was an ex-British vessel, specially built for shipment of nuclear materials. Designed to international standards specified by the International

Maritime Organisation, the vessel is double-hulled, rendering it damage resistant in any collision, crash or grounding situation. Also, it has sufficient stability to sustain itself afloat, preventing flooding in the holds, even if the hull is damaged. The cargo holds are protected by double bottoms running from forepeak to afterpeak and side tanks running the full length of the cargo area. Over 100 shipments of spent fuel have been safely completed by ships of this type. The ship was accompanied by an armed ship of the Japanese Maritime Safety Agency. In addition to the IAEA guidelines, the shipment complied with the Japan-US agreement for nuclear co-operation. Subsequent to the shipment, the USDOE conducted a study (19) of the safety of shipments of plutonium by sea and concluded that *"plutonium can be shipped safely by sea"*.

For accidents on the high seas, sinking in deep waters is the most likely type of severe accident that might befall sea shipments. The radiological consequences of this type of accident were assessed in the US report (19) which concluded that the risks to people and the biosphere are extremely small. There are recovery arrangements for accidents in shallow water.

Transportation of MOX fuels

MOX fuel assemblies generally contain between 5 and 10 per cent plutonium. This amount of fissile plutonium requires a Category I transportation system. The systems used to transport MOX fuel from the MOX plants (located in Dessel, Belgium; Cadarache and Marcoule, France; Sellafield, United Kingdom; and Tokai, Japan) to the German, Japanese, French, Swiss and Belgian reactors are in strict accordance with international agreements and IAEA recommendations. In particular, each transport system is aimed at ensuring the total integrity of the assemblies during transportation, and at decreasing, to the maximum extent possible, the exposure of operators during loading and unloading operations. MOX fuels are mainly transported by road, but sea and air transport can also be used.

Apart from transport of co-converted MOX powder within PNC Tokai works, there are three types of transport of MOX fuels regularly conducted within Japan. These are from the Tokai MOX fuel fabrication facility to:

- the experimental fast reactor Joyo;

- the heavy-water moderated Advanced Thermal Reactor Fugen (ATR Fugen); and

- the prototype fast breeder reactor Monju.

Summary

Plutonium has been safely transported, both nationally and internationally, for the last 35 years. The transport operations have been carried out in accordance with national and international law which has been based on regulations recommended by the International Atomic Energy Agency. This outstanding safety record is due to the design and development of the packaging and the rigorous checks that are carried out at every stage of the operation.

2.4   MOX fabrication

Historical milestones (20, 21)

Since the earliest days of the commercial utilisation of nuclear power, plutonium arising from reprocessing of spent fuel was recognised to be best used in FBRs. In the 1950s, the general opinion was, however, that reprocessing capacities in excess of the requirements for feeding FBR prototypes would be available up to a period of 20 years. It was then decided to launch an important R&D programme, which was conducted essentially within the framework of a co-operation agreement between EURATOM and the US Atomic Energy Commission (AEC).

The work sponsored by the USAEC was mainly performed at Hanford National Laboratory (plutonium utilisation programme) and Westinghouse (Saxton programme). The overall programme was aimed at developing all the necessary background data for the use of plutonium as mixed uranium-plutonium oxide fuel for LWRs, including the fabrication technology of such mixed uranium-plutonium oxide fuel.

The programme sponsored by EURATOM was performed by an association between Belgonucléaire and the Belgian National Laboratory (CEN/SCK) at Mol, Belgium. As in the US, the EURATOM programme covered all the facets of the technology, including the development of mixed oxide fuel fabrication. The MOX assembly loaded in BR3 (a 10 MWe PWR, which was in operation from 1962 to 1987) in 1963 constituted the world's first plutonium recycle in a LWR.

In the 1960s, interest in plutonium recycling grew steadily as plutonium surpluses appeared unavoidable. As a result, additional countries started R&D activities on MOX utilisation in LWRs: mainly Germany, and to a smaller extent, France, Switzerland, Italy and Sweden, as well as Japan for the ATR. The United Kingdom demonstrated MOX utilisation in the Windscale Advanced Gas-Cooled Reactor (AGR).

In the same period, mixed uranium-plutonium oxide, up to then investigated alongside to uranium-plutonium alloys, appeared also to be the most adequate fuel for the use of plutonium in FBRs. The development of mixed uranium-plutonium oxide fuel fabrication technology was therefore supported by FBR R&D activities, in particular in the US, France, Germany, Belgium and the UK.

Fabrication technology was, however, in its infancy during these periods: the main emphasis was placed on simplifying the manufacturing techniques and several alternative processes to pelletising-sintering (Figure 2) were tried out. The development schedule was successfully conducted at laboratory-scale, and already included small-scale demonstration irradiation programmes. It indicated quite clearly that alternative manufacturing techniques to pellets resulted in fuel which behaved perfectly, but which lacked the similarity to UO_2 fuel. Such a similarity was proven by the small-scale demonstration irradiations in LWRs to be a prerequisite for improved licensability. The pellet route was therefore adopted for all the pilot plants which started operation at the end of that decade: the CEA facility at Cadarache essentially for FBR fuel, the Belgonucléaire facility at Mol (1967) and the ALKEM facility at Karlsruhe (1965) for MOX and FBR fuel, and the UKAEA facility at Sellafield, which was unique in producing over 96 000 annular pellets for FBRs. The same decision was taken by PNC at Tokai-Mura (1965) for its pilot facility (PFDF) and the second facility (PFFF) for ATR and FBR fuels.

Figure 2. **Manufacturing techniques for plutonium-bearing rods**

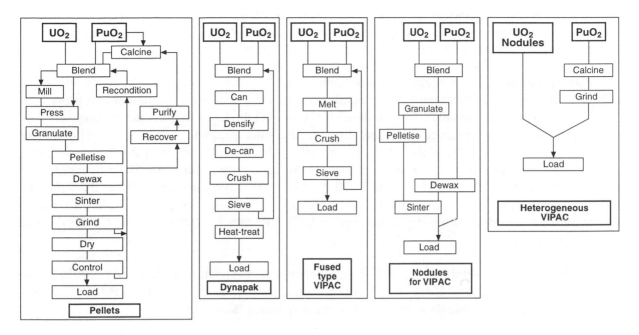

After the political decision of the US, in 1976, to indefinitely defer reprocessing, mixed uranium-plutonium oxide fuel technology was phased out in the US, but continued in:

- Germany and Belgium, both for MOX and FBR fuels;

- Japan, for ATR and FBR fuels; and

- France and UK, for FBR fuel.

The industrial mixed uranium-plutonium oxide manufacturing techniques utilised today were developed during that period, through a trial and error approach, based on lessons learned from the demonstration programmes. This was intrinsically and unavoidably a time consuming stage, as several years were required to accumulate fuel behaviour data under representative conditions (from the fabrication and irradiation points of view).

At the beginning of the seventies, the technology had reached enough maturity:

- Both to increase the capacity of the existing facilities to meet the then current demands of the respective domestic programmes:

  in France: CEA at Cadarache (known now as CFCa), for Superphénix FBR fuel (start of production in 1972); and

  in Japan: PNC (known as PFFF), for Fugen ATR and for Joyo FBR fuel (start of operation in 1972).

- And, to commission commercial mixed uranium-plutonium oxide fuel fabrication plants for both MOX and FBR fuels:

  in Belgium: Belgonucléaire (also known as P0) at Dessel (start of operation in 1973); and *in Germany:*   ALKEM (now known as SIEMENS BH) at Hanau (start of operation in 1972).

In the US, at the time of the political decision, in 1976, to indefinitely defer reprocessing, five organisations had licences enabling them to produce MOX fuel for LWRs on a pilot plant scale, with a cumulated annual production capacity of 50 to 70 t HM per year (22).

This period provided actually a reliable and strong foundation on which a viable industrial activity could be built. Progressively, remedial actions were initiated on the fabrication process to meet at least the same requirements as those for UO_2 fuel in terms of product quality, product performance and further reprocessing capability. In the latter particular concern, the reprocessing industry requested to assure single step dissolution of the mixed uranium-plutonium oxide fuel whatever its irradiation status. Results obtained from fuel manufactured by Belgonucléaire and ALKEM plants indicated that even after irradiation in normal operation conditions, the fuel was leaving a not acceptable level of undissolved residue after a single dissolution step in hot nitric acid representative of the dissolution step considered in UP2, UP3 and THORP plants (of COGEMA and BNFL, respectively).

To improve the solubility of the fuel, alternative manufacturing processes (MIMAS at Belgonucléaire, OCOM or AUPuC at SIEMENS) were developed, based on the manufacture of a master blend completely soluble in nitric acid after the sintering stage.

Regarding this particular request of the reprocessing industry, the consequences were deemed only minor for the fuel produced by CEA at Cadarache and by PNC at the Tokai plant (use of co-milling process of the whole fuel material).

The past thirteen years (1983 to 1996) were characterised by:

- the successful commercial operation of the COGEMA reprocessing plants at La Hague (UP2 and UP3);

- the start-up (in 1994) of the BNFL reprocessing plant at Sellafield (THORP);

- the successive delays in the large-scale deployment of the FBRs;

- the resulting start-up of large thermal MOX fuel utilisation programmes by the utilities; and

- the progress with fabrication plants, as described below.

Meeting the demand of the customers of those reprocessing plants resulted in a rapid expansion of the industrial utilisation of the MOX fuel which was limited, essentially, by the time required to implement and qualify new fabrication facilities, and by political considerations. Table 3 shows the status of the MOX fabrication plants in OECD countries. Within a period of about ten years (2005), a total capacity of 400 to 500 t HM per year will be operated, consistent with the total production of the then operating reprocessing plants.

In Russia, similarly to the programmes pursued in the OECD countries, plutonium has been considered as a nuclear fuel since the second half of the 1950s. The work has been essentially devoted to the development of fuel for fast reactors, as the fast breeder reactors using plutonium were considered as the most efficient way to expand nuclear power fuel resources. Initially, in 1957, a metallic alloy core was fabricated for the pulsing fast reactor IBR-30. Starting in 1959, mixed

Table 3   MOX fuel fabrication capacities – status

Country	Plant	Current* capacity (t HM/y)	Anticipated capacity (t HM/y)	Status	Remarks
1. Belgium	BN-Dessel P0 (FBFC-Dessel for mounting assemblies)	35 (since 1986)	40	Operation since 1973	–
2. France	COGEMA - CFCa (FBFC-Dessel for mounting assemblies, MELOX starting 1995)	15 (since 1988) 30 (at present)	35	Operation since 1962	Capacity for FBR fuel will be adapted to be able to produce as alternative up to 20 t HM
3. Germany	SIEMENS - Hanau	20 to 25 (before 1991)	–	Operation since 1972, but discontinued in 1991 by decision of local authorities	Should have been replaced by the 120 t HM/y facility (item 5). Now being decommissioned
4. France	MELOX - Marcoule	start-up	160	Operation start-up in 1995	–
5. Germany	SIEMENS - Hanau	–	(120)	Construction	Construction completed at 95%, but suspended by decision of local authorities. SIEMENS and German utilities decided not to operate it
6. UK	BNFL - MDF (Sellafield)	< 8	8	Operation since 1993	–
7. Belgium	BN-Dessel P1 (FBFC-Dessel for mounting assemblies)	–	40	Detailed layout completed	Construction to be decided
8. UK	BNFL - SMP (Sellafield)	–	120	Construction started in 1994	To be operational in late 1997
9. Japan	PNC - PFPF line (Tokai)	10	10	Operation since 1988	For FBR fuel
10. Japan	Yet unnamed	–	about 100	Preliminary conceptual design	Construction to be decided to be in operation shortly after 2000
11. France	MELOX Extension	–	50	Conceptual design completed	Construction decided. Start operation planned in 1999, full capacity planned by 2000

*   December 1996

uranium-plutonium oxide fuel was made, first for the BR5 and IBR-2 reactors, and later, from the mid-1970s, for the BOR-60 and for experimental sub-assemblies tested in the BN-350 and BN-600 reactors. Essentially, two technologies are being developed to process the plutonium (mainly low burn-up) into mixed uranium-plutonium fuel: pelletising and vibrocompacting. They are implemented at Mayak, Chelyabinsk and at RIAR, Dimitrovgrad, respectively. As a consequence of the delay of the construction of the next three to four units of the BN-800 fast power reactor, construction of a mixed uranium-plutonium oxide fuel fabrication plant, called COMPLEX-300, is currently suspended. The status of mixed uranium-plutonium oxide fabrication capacity in Russia is given in Table 4.

Small MOX fabrication facilities are also being operated in other countries, outside the OECD area, to fuel domestic power plants with demonstration or experimental MOX fuel (e.g. India).

Specific constraints for MOX fabrication (21, 1)

The predominant isotope, ^{239}Pu, is produced by neutron capture in ^{238}U. By successive neutron captures, ^{239}Pu can yield higher isotopes such as ^{240}Pu, ^{241}Pu and ^{242}Pu.

Small quantities of two other plutonium isotopes ^{236}Pu and ^{238}Pu are also produced during fuel irradiation by neutron capture in ^{235}U. These two isotopes are important because of their alpha, neutron and gamma activities or those of their daughter products.

This section addresses the consequences of these properties for the industrial implementation of the use of plutonium as fuel, in particular as MOX fuel for LWRs. The following items are considered:

- the radioactivity of the plutonium and the related constraints for MOX fuel fabrication and transportation; and

- the variability of the plutonium isotopic composition and the related constraints on MOX fuel reloads fabrication.

Radioactivity of plutonium

Plutonium (and its technical form PuO_2) is radioactive and toxic. It is essentially an alpha emitter, but also a producer of neutrons, X-rays, gamma-rays and beta particles. The radiotoxicity of plutonium requires that it is handled in air-tight boxes which are provided with plexiglas windows and with neoprene gloves (hence the name "glove boxes"). Since this radiation does not penetrate deeply, the radiological hazard is seriously reduced once the MOX is in the fuel rods.

The heat generated by the alpha activity originates essentially from ^{238}Pu. It results in PuO_2 transportation and storage problems which have found adequate solutions by proper design of transportation and storage systems, and in a progressive degradation of the fuel quality in the manufacturing process when $(U-Pu)O_2$ powders with their additives (lubricant, poreformer) are stored for some time.

Neutron activity (either directly from plutonium isotopes, or from (α, n) reactions mainly with oxygen) is practically constant with time and depends mainly on the isotopic composition of the plutonium. Proper shielding has to be implemented at every stage of the MOX fuel cycle, while avoiding this becoming a limiting feature for LWR fuel.

Table 4 Mixed uranium plutonium oxide fuel fabrication facilities in Russia

Facilities	Fabrication process	Reactors	Fabrication experience (t MOX)	Annual capacity
PAKET at Mayak, Chelyabinsk	UO_2-PuO_2 mechanical mixing and pelletising	BR5 BN-350 BN-600	3 800 rods : 1 t 1 778 rods $\}$ 1 t 1 524 rods	Up to 1996: 300 kg MOX for 10 to 12 FAs After 1996: 1 t MOX for 30 to 40 FAs
RIAR at Dimitrovgrad	Vibro-packed fuel	BOR-60 BN-350 BN-600	12 800 rods $\}$ 254 rods $\}$ 1.8 t 762 rods	1 tonne MOX
Complex-300 at Chelyabinsk	Masterblend of 30 w/o Pu/HM by mechanical blending of PuO_2 and UO_2. Secondary blending, precompaction, granulation, pelletising	BN-600 BN-800 VVER-1000 (depending on evaluation underway)	–	60 t HM. Construction is 50% complete and currently suspended

The weak gamma activity builds up continuously after the last purification step of the PuO_2 at the reprocessing plant: it is generated by the decay of ^{241}Pu into ^{241}Am (4.7 per cent of ^{241}Pu decays per year) leading not only to radiological problems, but also to a reduction of the fissile plutonium inventory. This major gamma radiation contributor is relatively easy to shield, since it is at a low energy level; the radiological problems stem essentially from dust depositing on the equipment and the internal surfaces of the glove boxes.

Most of the high energy radiation arises from the decay of ^{236}Pu eventually into ^{212}Bi and ^{208}Tl through the ^{232}U decay chain. This source is, however, much smaller than the ^{241}Am and builds up at a five times slower rate. For all practical purposes in PuO_2 handling and in the front-end of MOX fuel manufacturing, the ^{241}Am content is nowadays the main gamma dose contributor. As a result, incorporating the plutonium into MOX fuel requires effective gamma shielding. This has been confirmed by dose rate measurements during fabrication campaigns in various facilities.

As plutonium is usually transferred from the reprocessor to the fuel manufacturer as PuO_2 powder, personnel at the manufacturing plant (mainly the maintenance personnel in an automatised plant) will receive important doses at the fabrication stages of powder mixing and pelletising, while the dose rates measured when manufacturing fuel assemblies are two orders of magnitude lower. After prolonged storage, it will become economically impracticable and socially unacceptable to utilise the plutonium without americium removal in a manually operated fuel fabrication plant. The limiting storage period depends on the plutonium composition, on whether plutonium is stored as PuO_2, primary blend (if possible), MOX pellets, fuel rods or fuel assemblies and on the equipment of the plant.

Practical storage limits are, respectively, two to three years (present manually operated manufacturing plants) to five to six years (design basis for future plants) for a typical LWR PuO_2 feed, and 10 years for MOX fuel rods.

If PuO_2 is stored for longer periods before proceeding with fuel fabrication, an extra processing step is required, namely the elimination of ^{241}Am. It consists of the following sequence of operations: dissolution, purification (i.e. striping of the americium) and reconversion to PuO_2 or codenitrated $(U-Pu)O_2$, if this is the feed material (Japanese fabrication facilities).

The cost of such reconditioning, being a simplified reprocessing operation, may be an important economic penalty on the use of plutonium in LWRs, as well as in FBRs. It is the reason why plutonium should not be stored as PuO_2 powder as delivered by the reprocessing plant but should instead be manufactured into MOX fuel rods or even assemblies as soon as possible. The fuel rods can safely be stored at the fuel fabrication plant and the assemblies in the power plant dry storage vaults or spent fuel pools.

Variability of the plutonium isotopic composition

a)   The isotopic composition of plutonium contained in MOX fuel at the time of loading in the reactor may deviate, sometimes significantly, from the isotopic composition of plutonium contained in spent uranium fuel that the utility had sent to the reprocessor for recuperation of plutonium for use in the manufacture of that MOX fuel.

1) The isotopic composition of the plutonium contained in the spent uranium fuel assemblies at discharge after irradiation depends mainly:

- on the initial enrichment of the uranium in ^{235}U: this is usually uniform for equilibrium reloads, but varies in transition cores (initial cores, extension of discharge burn-ups and/or cycle lengths); and

- on the burn-up of each fuel assembly at discharge: some variability already arises from the dispersion of discharge burn-ups within a batch and from one discharge batch to the next.

It also depends, to some extent, on the detailed power rating history of each fuel assembly for a given discharge burn-up.

2) Isotopic composition of the plutonium arising from the reprocessing of spent uranium fuel assemblies depends also on the cooling time between the end of the irradiation of the spent fuel assemblies and the reprocessing and chemical separation of the plutonium, and on the storage time of the PuO_2.

Indeed, when the spent fuel is removed from a reactor, fresh plutonium production ceases immediately and radioactive decay becomes the dominant feature. The half-lives of ^{238}Pu, ^{239}Pu, ^{240}Pu and ^{242}Pu are sufficiently long, so they remain effectively unchanged during that cooling period of five to ten years. The shortest-lived isotope of plutonium, ^{241}Pu, decays with a half-life of 14.4 years and is replaced by the neutron absorbing (and gamma emitting) ^{241}Am.

3) The plutonium returned to the utility after the reprocessing of the spent fuel of that utility is not necessarily the plutonium coming from its reprocessed spent fuel. It depends on the contractual arrangements between the reprocessor and the utility:

- In the case of the THORP facility at Sellafield, as well as the UP3 facility at La Hague, any of the plutonium produced during the first ten years of operation is allocated each year, in terms of fissile plutonium, to each utility of the base load customers contract in proportion to the theoretical quantity of fissile plutonium in the spent fuel assemblies the utility has committed to reprocess in the facility during that period.

  Each utility is, therefore, being delivered by the reprocessor a given fraction of the total plutonium produced each year by the plant, without consideration of the isotopic composition of the plutonium contained in the spent fuel assemblies sent to the reprocessor by that utility.

  The utilities, base load customers, must agree among themselves on the sequence of plutonium allocation to each other, such that each utility would be able to make available in due time, to the MOX fuel manufacturer, the quantity of plutonium needed for manufacturing the contracted quantity of MOX fuel assemblies and meeting the requirements of maximum ^{241}Am content.

- It might happen that a utility needs an additional quantity of plutonium to reach the quantity needed for the manufacture of MOX fuel assemblies. The reprocessor is, sometimes, able to provide that additional quantity in advance of the plutonium to be obtained later from the reprocessing of spent fuel.

That additional plutonium might come from higher or lower burn-up fuel than the bulk of the plutonium allocated to the utility customer. The isotopic composition of that additional plutonium provided in advance might therefore deviate significantly from the isotopic composition of the bulk of the allocated plutonium.

b) The isotopic composition of the plutonium received at the manufacturing plant might be significantly different from one batch of PuO_2 to another. Even the plutonium batch made available by one utility for one fabrication campaign may consist of PuO_2 lots of quite different isotopic compositions.

c) The MOX fuel manufacturing plants themselves must face three main constraints in the organisation of a fabrication campaign:

1) The plutonium lots with the highest [241]Am content must be used first in order to reduce personnel exposure in the fabrication plant, whatever the specific fabrication campaign for which that plutonium has been normally delivered.

In that regard, each fabrication campaign must incorporate, as soon as possible, the green or sintered pellet scraps originated from the previous fabrication campaigns to avoid americium build-up.

2) The engineers in charge of MOX fuel nuclear design and core management require a MOX fuel as homogeneous as possible with regard to the plutonium isotopic composition in order to be able to consider all the fuel rods of one plutonium content interchangeable in the same fabrication campaign. If this is not the case, the position of each MOX rod in each assembly must be determined and controlled or additional engineering hot spot factors must be taken into account in design and licensing.

3) The manufacturer must provide the utility an amount of plutonium in the form of MOX fuel equivalent (from the energy content point of view) to the amount received from the reprocessor in the form of PuO_2, while meeting the strict and accurate safeguards obligations imposed by the national and international authorities.

Actual operating manufacturing plants

The process used in MOX fuel fabrication differs from that used in UO_2 fuel fabrication, mainly at the level of the preparation of the powders (Figure 3).

The steps involved in the preparation of the MOX powders, equivalent to the enrichment step of the [235]U in uranium for the UO_2 fuel, are those which require the largest protection for the fabrication plant operators due to the radiotoxicity of the PuO_2.

Figure 3. Actual MOX fabrication flow sheets

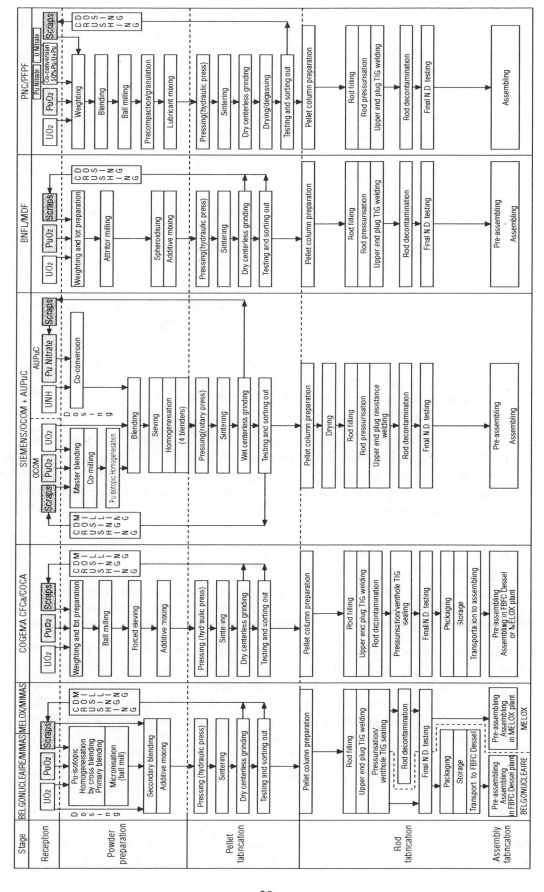

The other fabrication steps (granulation, if requested by the process, pressing, sintering, grinding, rod filling and welding) are essentially similar to equivalent steps used in the fabrication of UO_2 fuel. But, taking into account the radiotoxicity of plutonium, these steps require fabrication inside the glove boxes.

The greatest attention is paid to the protection of the operators against radiation exposure, from external radiation fields or from the intake of radioactive material, and avoiding dispersion and release of plutonium aerosols from the plant. The following principles are applied:

- Plutonium-bearing materials are strictly confined inside air-tight glove boxes, under constant under-pressure.

- All glove boxes are located within fire proof zones, also with tight barriers. The plant exhaust system provides for a stepwise decrease of the air pressure in the direction of the areas with higher contamination risks. The atmosphere in all zones is exchanged several times per hour.

- There are separate ventilation systems for each glove box line and working areas. The air exhausted from the boxes and all working areas passes through several absolute filters connected in series before leaving the building, preventing the release into the atmosphere of alpha-contaminated materials.

- The atmosphere inside the working rooms and all streams of the plant are continuously monitored for contamination. To ensure early detection of radioactivity dispersion, all equipment, as well as hands, feet, and clothing of each worker are permanently checked: no contamination is allowed outside the glove boxes.

- The operation has been gradually upgraded from initially pure manual lines to later on mechanical hands-off operation. Shielding was progressively incorporated to protect the operators in the most exposed areas. During operation of the facilities, the main radiological hazard, was, as expected, found to result from plutonium dust deposited on the surfaces inside the glove boxes.

- The reduction of personnel doses can be achieved by keeping glove boxes clean, by using compact equipment and tight fittings during the transfer of MOX materials from one equipment to another, as well as by avoiding the hold-up of those materials.

- Criticality safety has been achieved by selection of safe geometry, limitation of moderating material, and, in case of non-standard operations, careful handling by monitoring the masses involved.

- Mounting the fuel rods in assemblies is based on the same operations as for UO_2 fuel assemblies. Special care is taken to reduce the doses (neutrons and gammas) received by the personnel involved in the fuel assembly inspection and to guarantee the correct loading pattern of the fuel rods in the assemblies (indeed, in a PWR fuel assembly the MOX fuel rods are distributed according to three different plutonium contents, while the uranium enrichment is uniform in the UO_2 fuel assembly).

Belgonucléaire Dessel P0 plant (23, 24, 25, 26, 27)

a) Background

The plant is located next door to the UO_2 fuel fabrication plant of FBFC, on the Mol-Dessel nuclear site, which houses also the domestic waste conditioning and storage facilities (Belgoprocess), the national laboratory (CEN/SCK) and the joint European Bureau of Nuclear Measurements (EURATOM/Geel).

Operational since 1973, its earlier fabrication process was used for the partial reloads and fuel supplies for the following thermal power reactors and facilities: Garigliano (Italy), BR3 and Venus (Belgium), CNA-Chooz (France), Dodewaard (Netherlands), Oskarshamn (Sweden) and NPD (Canada). It was also applied to fuel supplies for the following fast breeder reactors and facilities: SNR and KNK (Germany), Rapsodie and Phénix (France), and DFR and PFR (UK).

These first ten years of operation have laid down the bases for all modifications or improvements in the fields of fuel fabrication and control processes, waste management, safety and safeguards, which were implemented in the 1984 refurbishment. On this occasion, the capacity of the plant has been upgraded to 35 t HM per year of standardized MOX fuel, level achieved consistently since 1989.

Belgonucléaire-Dessel had gained the experience of mounting plutonium-bearing rods in assemblies for LWRs (BR3, CNA-Chooz) and for FBRs (40 per cent of SNR first core). However, with the industrial implementation of MOX fuel manufacture starting from the mid-1980s, mounting the MOX fuel rods in bundles has been transferred to FBFC-Dessel, taking advantage of the large (30 years) experience of FBFC-Dessel, where more than 15 000 LWR fuel assemblies of several designs have been mounted up to the end of 1995, of which 700 are MOX fuel assemblies (PWR and BWR).

b) MOX fabrication process

The actual MOX fuel fabrication process, implemented in 1985, known as the MIMAS process is the most recent evolution step of the four successive fabrication techniques (see Table 5) developed by Belgonucléaire to produce MOX fuel characterised by an intimate dispersion of the plutonium in the fuel matrix. MIMAS is a proprietary acronym derived from "MIcromized MAster blend", a key intermediate product in the fabrication process. The MIMAS process is schematically shown in Figure 3: one line of production of primary and secondary blends is feeding two lines for pellet and rod production.

The fresh PuO_2 powder is supplied by the utility according to Belgonucléaire specifications (in fact, in agreement with the characteristics of the PuO_2 produced by COGEMA). The maximum average [241]Am content should not exceed 17 000 ppm when used in the fabrication.

The fresh uranium powder is supplied by the utility as depleted UF_6 (essentially tails from enrichment plant) and is converted into free flowing UO_2 by Belgonucléaire's qualified subcontractors (namely ex-AUC conversion or ex-ADU TU2 conversion process).

All the scrap (*i.e.* material processed in excess, material rejected by quality control and the residues from the dry grinding operation) are systematically recycled within the process by micronisation as for primary blends and by re-introduction in the primary blends (mainly) or in the secondary blends.

The plutonium containing liquid solution, as a result of the chemical analysis, is chemically treated to precipitate the plutonium as PuO_2 which is recycled in the process.

Table 5 **Evolution of the MOX fuel fabrication techniques of Belgonucléaire**

Fuel type	Period	Advantages and inconveniences
Granulated (UO_2 +PuO_2) blend	1960-1962 and 1965-1969 (laboratory)	Assumed best similarity to UOX fuel. Contamination levels, personnel exposure and waste arisings resulting from complex handling of fine powder.
PuO_2 blended into granulated UO_2	1967-1975 (pilot facility)	Significant departures from UO_2 fuel behaviour. Simplified handling of fine powder. Large plutonium-rich agglomerates. Unfavourable thermal conductivity. High fission gas release.
"Reference", *i.e.* PuO_2 blended into free flowing UO_2	1973-1984 (fabrication plant)	Fuel microstructure governed by the UO_2 matrix microstructure. Occurrence of plutonium-rich agglomerates. Too large proportion of the plutonium in insoluble residues (reprocessing problem).
"MIMAS", *i.e.* mixing of free flowing UO_2 and a micromised (UO_2 + PuO_2) primary blend.	Since 1983 (fabrication plant)	Same advantages as the reference MOX and applicability of its data base. Disappearance of the plutonium-rich agglomerates issue due to dilution. Resolution of the reprocessing issue.

The process wastes are treated as plutonium contaminated materials: they are sorted depending on the type and contamination level and transported to be conditioned and packaged for storage and disposal by volume reduction and encapsulation.

c)   Capabilities and operating experience

MOX fuel manufactured in the Dessel site can be either PWR (14 x 14, 16 x 16, 17 x 17, 18 x 18) or BWR (8 x 8, 9 x 9, 10 x 10) type fuel. The reference production capacity is, typically, 35 t HM per year, on the basis of two fabrication campaigns of 17 x 17 PWR fuel, with three different plutonium contents per assembly.

The actual capacity is, however, affected by factors such as :

- the quantity of MOX fuel to be manufactured per campaign;
- the number of different plutonium contents per fuel assembly;

- the number of different fuel rod designs; and

- the plutonium content in the MOX fuel (although manufacture of MIMAS fuel has been experienced up to 14 w/o Pu/HM, it is limited to 10 w/o due to the limitation by licence in FBFC-Dessel).

From 1973 till the end of 1995, the mixed uranium-plutonium oxide fabrication facilities of Belgonucléaire have processed about 16 t Pu(t) into 290 t HM contained in 180 000 fuel rods and 970 fuel assemblies for both thermal and fast reactors. The processed plutonium has had a ^{240}Pu content from about 5 per cent in plutonium, and an ^{241}Am content up to about 20 000 ppm in plutonium.

Out of these quantities, about 280 t HM were processed into MOX fuel and used to fill about 160 000 fuel rods loaded in 900 fuel assemblies.

As far as the actual MIMAS product is concerned, about 244 t HM have been used in about 136 000 MOX rods, loaded in 615 MOX assemblies for use in commercial LWRs located in France (26 reloads in seven reactors), Germany (four reloads in four reactors), Switzerland (five reloads in one reactor) and Belgium (two reloads in two reactors).

As a result of the decreasing quality of the PuO_2 and of the increasing quantities of PuO_2 being processed, programmes are continuously conducted aiming at decreasing the occupational doses of the personnel. Typical actions are:

- adaptation of working procedures;

- dust collection and reduction of the size of the glove boxes;

- introduction of gamma and neutron shieldings on the fabrication line in order to decrease significantly the background radiation in the fabrication halls; and

- progressive automation of the working stations.

Collective doses per tonne HM produced have been reduced from about 130 mSv in 1986 to about 40 mSv in 1994, with, in 1994, an average individual value around 5 mSv/y for the workers of the most exposed areas.

Similarly, a programme was initiated to identify the waste generating operations with a view to reducing them and to make the personnel concerned with the problem. The result of these actions gives evidence that the waste production is perfectly under control, with plutonium contained in the waste being less than 0.1 per cent of the plutonium in the delivered MOX fuel.

d) *Assembling operations*

All the MOX rods produced by Belgonucléaire are mounted in fuel bundles/assemblies in the FBFC-Dessel plant. A new assembling workshop, adjacent to the existing UO_2 assembling workshop, has been constructed taking into account the severe constraints also considered for the P1 plant. The new workshop comprises three halls, respectively, a fully bunkered storage room for MOX fuel assemblies, a hall for assembling operations, and a hall for receiving the MOX rods from Belgonucléaire and for dispatching the MOX assemblies using the present transport containers and those anticipated in the future.

The production capacity is consistent with the cumulative capacity anticipated for the P0 and P1 Belgonucléaire plants. The operations licence and the construction permit were granted in 1993. The new storage hall has been in use since 1995, and the new assembling workshop is to be fully operational in 1997.

COGEMA Cadarache plant (CFCa) (23, 28, 29)

a) Background

The CFCa (Centre de Fabrication de Cadarache) has been processing plutonium fuel since 1962, mainly for fast breeder reactors. It started with the fabrication of oxide fuel for Rapsodie on a two kg per day capacity line, then it increased continuously with successively metal fuel for experimental reactors (five kg PuAl per day), oxide fuel for Phénix (20 kg oxide per day), then for Superphénix (up to 120 kg oxide per day).

Since 1989, the Superphénix line has been converted to MOX fuel fabrication, with a 20 t HM per year reference capacity.

Originally, the facility was part of the installations operated by the CEA Cadarache. However, in view of the continuous development of the CFCa and its increasing industrial objectives, the facility has been transferred to COGEMA which is thus operating the CFCa since 1 February 1991.

All the plutonium-bearing rods manufactured within the framework of the fast reactors projects are mounted in fuel bundles inside the CFCa facilities. However, as for the MOX rods manufactured by Belgonucléaire-Dessel, the MOX rods manufactured by CFCa are transported to FBFC-Dessel for their mounting in fuel bundles. With the start-up of MELOX (see below) the MOX rods can also be transported to MELOX.

b) MOX fabrication process

The fabrication process to manufacture MOX fuel is based on the large experience gained with the fabrication of FBR (U-Pu)O_2 fuel: since 1962, about 135 t HM were processed in the facility. A specific route, called COCA (Cobroyage Cadarache) developed for FBR fuel has been adapted for LWR fuel. The COCA fabrication process is outlined in Figure 3.

The starting materials are:

- PuO_2 from COGEMA or from BNFL reprocessing plants. The maximum [241]Am content shall not exceed 15 000 ppm in plutonium when used in fabrication. The [240]Pu content must be higher than 17 per cent w/o in plutonium.

- UO_2 prepared according to the ADU type conversion process (in COGEMA's TU2 plant).

Besides the usual recycling, part of the scraps are recycled through a classical chemical treatment. Part of the process is common to the treatment of analytical solutions used for chemical tests and destructive control operations. A new specific dry recycling installation will be put in operation in order to shorten the duration and to reduce the cost of fuel materials recovery: the scraps – after calcination in a reducing atmosphere – will be ball milled; such re-activitated (U-Pu)O_2 powder may be added to fresh powder up to 20 per cent of a blender load.

c) Capabilities and operating experience

MOX fuel manufactured at the CFCa site can be of PWR type fuel (16 x 16, 17 x 17 or 18 x 18) or BWR type fuel. The production capacity reached in 1995 was about 30 t HM per year on the basis of two fabrication campaigns of 17 x 17 PWR fuel with three different plutonium contents per assembly.

The Pu/HM is limited to 12.5 w/o, with uranium being natural or depleted (tails from the enrichment plants).

The capacity of the CFCa is now being progressively increased to reach 35 t HM per year starting in 1996. Such capacity will be implemented in two lines:

- *line 1*: with 20 t HM per year where the two processes MIMAS and COCA will be made possible, in particular MIMAS or COCA for use in MOX fuel (LWRs) and COCA for use in FR fuel (in equivalent capacities); and

- *line 2*: with 15 t HM per year, based on the COCA process for use for MOX fuel or the equivalent capacity for FR fuel.

From 1962 to the end of 1994, about 27 t of plutonium have been processed in about 158 t HM, in about 470 000 fuel rods, contained in about 2 200 fuel assemblies.

As far as the actual COCA LWR-MOX fuel is concerned, about 15.7 t HM have been used in 26 000 MOX rods, loaded in 100 MOX assemblies for use in EDF's 900 MWe PWRs.

The quality of the plutonium delivered to CFCa (from COGEMA reprocessing plants) follows the same trends as in Belgonucléaire-Dessel.

The records show that no more than 5 per cent of the staff receive an annual dose of more than 20 mSv.

SIEMENS – Hanau plant (30, 31, 32)

a) Background

Processing of plutonium into mixed uranium-plutonium oxide fuel by the SIEMENS Hanau Fuel Fabrication Facility, formally known as ALKEM, was first handled in a production line at Kernforschungszentrum Karlsruhe (KfK) from 1965 to 1971 and was then transferred to two production lines in Hanau, which began operation in 1972, based on direct blending of PuO_2 powder with free flowing UO_2 powder (similar to the reference process of Belgonucléaire - Table 5).

The original purpose of ALKEM was, beside performing R&D work, to convert plutonium nitrate from the small Reprocessing Facility Karlsruhe, as well as PuO_2 from French reprocessing plants, into fuel for the German fast reactors programmes (KNK, SNR-300).

At a rather early stage the recycling of plutonium produced in thermal reactors became attractive and, since the mid-1980s, the fabrication of MOX fuel from PuO_2 received from French and UK reprocessing plants became the primary business operation.

b)   MOX fabrication process

To meet the reprocessability requirements, two new processes (see Figure 3) were developed and implemented in the late seventies:

- The OCOM process (Optimised CO-Milling): homogeneous distribution is achieved by intensive milling of a mixture of PuO_2 and UO_2. This process was developed for the processing of PuO_2 delivered by French and British reprocessing plants. It is very similar to the MIMAS process of Belgonucléaire and is used with plutonium oxide powder as the starting material.

- The AUPuC process (Ammonium-Uranyl-Plutonyl-Carbonate): using the precipitation of the AUPuC complex by adding ammonia and carbon dioxide to a solution of uranyl nitrate and plutonyl nitrate, followed by a calcination, a high degree of homogeneity is achieved. This coprecipitation process is used if plutonium is delivered in the form of a Pu(IV) nitrate solution.

The dry scrap materials are recycled within the process by ball milling as for the masterblends followed by re-introduction, either in the masterblends, or directly during the preparation of the press feed by blending.

The plutonium contained in the wet wastes and in the process wastes has to be dissolved in a hydrofluoric acid dissolving station, and recovering is performed through the AUPuC conversion process.

c)   Capabilities, operating experience and actual status

Mixed uranium-plutonium oxide fuel manufactured in SIEMENS Hanau plant can be either PWR type, BWR type or FBR type fuel.

Although for MOX fuel production, the reference capacity was set at 35 t HM, the capacity during the period 1987 to 1991 was between 20 and 25 t HM of MOX fuel per year, essentially based on the OCOM process, and on a smaller basis, via the AUPuC process (this, in a 150 kg plutonium per year facility, from plutonium nitrate, *i.e.* from scrap recovery).

The throughput accumulated from 1966 to mid-1991 is as follows:

- *In terms of Pu fissile*:   5 817 kg in total, shared between:
  - MOX fuel for FRs: 1 357 kg; and
  - MOX fuel for LWRs: 4 460 kg.

- *In terms of Heavy Metal*:   163 862 kg in total, shared between:
  - MOX fuel for FRs: 5 883 kg; and
  - MOX fuel for LWRs: 158 029 kg

with a maximum of FR fuel in the years 1981 to 1985, and an increasing throughput of LWR MOX fuel from about 5 t HM per year in the 1970s to more than 20 t HM per year in the period 1987 to 1991. The processed plutonium has experienced [240]Pu and [241]Am contents ranging from about 5 per cent and up to 10 000 ppm in the plutonium, respectively.

Notwithstanding increasing radiation source strength and increasing throughput, the individual dose of the workers, with a few exceptions, did not exceed 10 mSv per year. The mean dose of all exposed persons always layed in the region of 5 mSv per year.

On 19 June 1991 a contamination incident occurred in the central storage facility, caused by a leak in the double foil covering of a can containing MOX powder. Five persons were contaminated internally, a yearly effective dose of 38 μSv resulted for the highest exposed person, this is 0.075 per cent of the maximum permitted value. The whole plant was shut down by the Hesse Ministry of Environment and despite the fact that an overpack was developed to avoid future damaging of foils, restart has not been permitted. The plant is now emptied from plutonium. SIEMENS has decided not to operate this plant any more.

MDF facility of BNFL – Sellafield (33, 34, 35)

 a)   Background

BNFL and UKAEA have collaborated over the last 30 years in the manufacture of plutonium fuels. In the early 1960s a total of about three tonnes of MOX fuel was manufactured for a variety of reactor systems including PWR, BWR and gas-cooled reactors, using small-scale equipment. Between 1970 and 1988, approximately 18 t of fast reactor fuel was manufactured in a purposely built production scale plant at enrichments of up to 33 per cent. Additionally, small quantities of experimental fuel were made in small-scale facilities. The importance of fuel homogeneity was recognised, satisfactory standards were achieved and fuel performed well in reactors. This experience, together with associated development work, has formed the foundation of the more recent development with capability to manufacture MOX fuel for LWRs.

Against this background and the increased commercial interest worldwide in MOX fuels, BNFL and UKAEA have collaborated over the last few years to develop facilities for manufacture of plutonium-bearing fuels. BNFL has recently taken over UKAEA's interest in the collaboration including the development facilities and is currently operating a small-scale manufacturing facility, the MOX Demonstration Facility (MDF) at Sellafield.

This facility was created by modifying some existing UKAEA facilities at Sellafield. The project commenced in early 1990 and the plant became operational in 1993, and is now manufacturing fuel for irradiation in a PWR. The design capacity is a nominal 8 t HM per year, providing a manufacturing facility for fuel assemblies for commercial LWRs. The plant is currently equipped to manufacture only PWR design fuel, but is capable of modification to extend the range of fuels manufactured to include a variety of pellet, fuel rod and assembly designs, such as those used in BWRs. These modifications will be introduced in line with customer requirements.

 b)   MOX fabrication process

The flowsheet used in the MDF for MOX fuel manufacturing (Figure 3) is similar to that used by other manufacturers with the exception of the introduction of the BNFL developed "short binderless route". This makes use of a high energy attritor mill to blend and homogenise the feed powders, followed by conditioning in a spheroidiser to provide free flowing press feed granules.

The MDF has been designed to handle plutonium originating from LWR uranium fuels reprocessed in THORP. Preference is given to freshly separated plutonium to avoid the additional dose arising from in-growth of americium. Uranium feedstock will normally be depleted or natural uranium dioxide manufactured by the BNFL developed "integrated dry route" process.

The plant and processes are designed to minimise any environmental impact. To the maximum extent possible, fuel residues are recycled within the process. Other process wastes are treated as Plutonium Contaminated Material (PCM) and will be conditioned and packaged for storage and disposal by volume reduction and encapsulation in concrete.

c) Capabilities and operating experience

The plant is designed to handle PuO_2 which has been separated from reference LWR uranium fuels irradiated to 30 GWd/t, cooled for five years before reprocessing, and aged up to one year before conversion into MOX fuel.

The nominal design capacity of the plant is 8 t HM per year, although the actual capacity will be affected by the particular designs being made.

MOX manufacture commenced in 1993 and the first four assemblies for commercial reactor loading have been delivered in 1994. Over the next twelve months of operation the production rate has been steadily increased as operating experience was gained. The operating life of MDF will be reviewed when the larger-scale Sellafield MOX Plant (SMP) has been made operational.

PNC – Tokai (36)

a) Background

There are currently three plutonium fuel fabrication facilities in Japan, all located in the same area as the Tokai reprocessing plant, owned and operated by Power Reactor and Nuclear Fuel Development Corporation (PNC), a governmental organisation.

These three facilities are designated as follows:

facility No. 1:   PFDF (Plutonium Fuel Development Facility);

facility No. 2:   PFFF (Plutonium Fuel Fabrication Facility); and

facility No. 3:   PFPF (Plutonium Fuel Production Facility).

Table 6 summarises their start of operation, the purpose of their operation and their actual status. These facilities produce, in particular, the plutonium bearing fuel for the fast reactors Joyo and Monju, and the MOX fuel for the Fugen ATR. About five tonnes of plutonium have been processed to produce 144 tonnes MOX fuel in the last 15 years.

Table 6 PNC plutonium fuel fabrication facilities

Facility	Completion of construction	Start of operation	Purpose of operation	Actual status	Maximum capacity of Pu handled	Maximum Pu content	Limiting Pu Isotopic composition
No. 1 PFDF (R&D)	November 1965	January 1966	Basic research. Fabrication of fuel for irradiation experiments	In operation	300 kg Pu annually	35%	Criticality safety assessment: 100% ^{239}Pu. ^{241}Am: 10 000 ppm. Radiation exposure: 50 mSv/y
No. 2 PFFF Line 1 (FR line)	August 1969	1973	Fabrication of Joyo driver fuels (Joyo: experimental fast reactor having a thermal output of 100 MW, operated by PNC)	Interrupted since 1987	–	30%	Criticality safety assessment: 100% ^{239}Pu. ^{241}Am: 10 000 ppm. Radiation exposure: 50 mSv/y
Line 2 (ATR line)	September 1970	March 1972	Fabrication of Fugen driver fuels (Fugen: advanced thermal reactor, heavy-water for moderation, light-water for cooling, having outputs of 515 MW thermal and 165 MWe, operated by PNC). Fabrication of very small quantities of MOX fuel for BWRs (irradiation tests)	In operation	850 kg Pu annually	2.7% for Fugen 5.7% for BWR-MOX	Criticality safety assessment: 100% ^{239}Pu. ^{241}Am: 10 000 ppm. Radiation exposure: 50 mSv/y
No. 3 PFPF (FR line)	October 1987	October 1988	Fabrication of Monju and Joyo driver fuels (Monju: fast breeder reactor having an output of 714 MWt and 280 MWe, operated by PNC)	In operation	2.5 t Pu annually	32% for Monju	Criticality safety assessment: 80% ^{239}Pu , 10% ^{240}Pu, 10% ^{241}Pu. ^{241}Am: 30 000 ppm. Radiation exposure: 50 mSv/y

b)   MOX fabrication process

The reference process (Figure 3) has been established with the following starting materials:

- UO_2 received from usual $UF_6 - UO_2$ convertors; and

- a co-converted $(U-Pu)O_2$ powder obtained by microwave codenitration of mixed solution of plutonium nitrate and uranium nitrate (Pu/U ratio = 1), outputs from the PNC reprocessing plant.

PNC is, however, able to receive the PuO_2 powder which is the usual output from the European reprocessors.

The co-converted $(U-Pu)O_2$ powder, or the PuO_2 powder, is mixed with the UO_2 powder at the plutonium content specified for the final product. The powder is then ball milled (use of a conventional ball mill), precompacted and granulated. Clean pellet scraps are crushed into powder which is recovered and then recycled to the head-end of the process as feed material. Clean powder scraps are sintered once and then treated in the same way as the clean pellet scraps. Dirty scraps are stored and are to be processed with purification. Wastes are treated with volume reduction process and/or incineration process to reduce their volume and, then, temporarily stored.

COGEMA – MELOX (23, 27, 29, 37)

a)   Background

The facility is located in the south of France, near Avignon, on a 50 000 square meters area on the site occupied by COGEMA at Marcoule, the cradle of the French nuclear industry.

Whenever possible, the fabrication processes have been based on techniques already used and qualified in industrial scale plants.

In that respect, the MELOX process and technology are, on the whole, relatively close to proven processes and technology already existing in the Belgonucléaire Dessel MOX plant (essentially for MOX powder preparation and pellets) and in the FBFC UO_2 plants at Dessel, Roman and Pierrelatte (essentially for MOX rods and MOX assemblies), see Figures 3 and 4.

The objectives set up for MELOX, from the start, concerning radioactive contents or characteristics of nuclear material and regarding radiation protection, resulted in more automation than the corresponding operations at Belgonucléaire and FBFC.

Severe constraints had to be dealt with ensuring an aseismic design of VIII-IX degree intensity level on the Mercalli scale. Consequences on civil works and mechanical structures such as glove boxes, pipes, air ducts and cable supports were heavy, increasing the overall MELOX complexity.

The main building houses the production line on two separate floors, from UO_2 and PuO_2 powders reception and storage devices to fuel assemblies storage and conditioning for transport. Based on a production of 100 t HM per year, the plant capacity is of the order of 100 000 pellets or 350 rods a day, a daily production of about one fuel assembly. Thanks to main equipment characteristics, the general layout of the MELOX production building will support an increase of nominal production (provided authorisation to increase capacity is obtained). With three-shift operation and limited additional equipment installation (in particular an additional sintering furnace), MELOX capacity could reach 160 t HM per year in the future, *i.e.* 60 per cent more than actual licensed characteristics.

Figure 4. **MELOX: main and new flows**

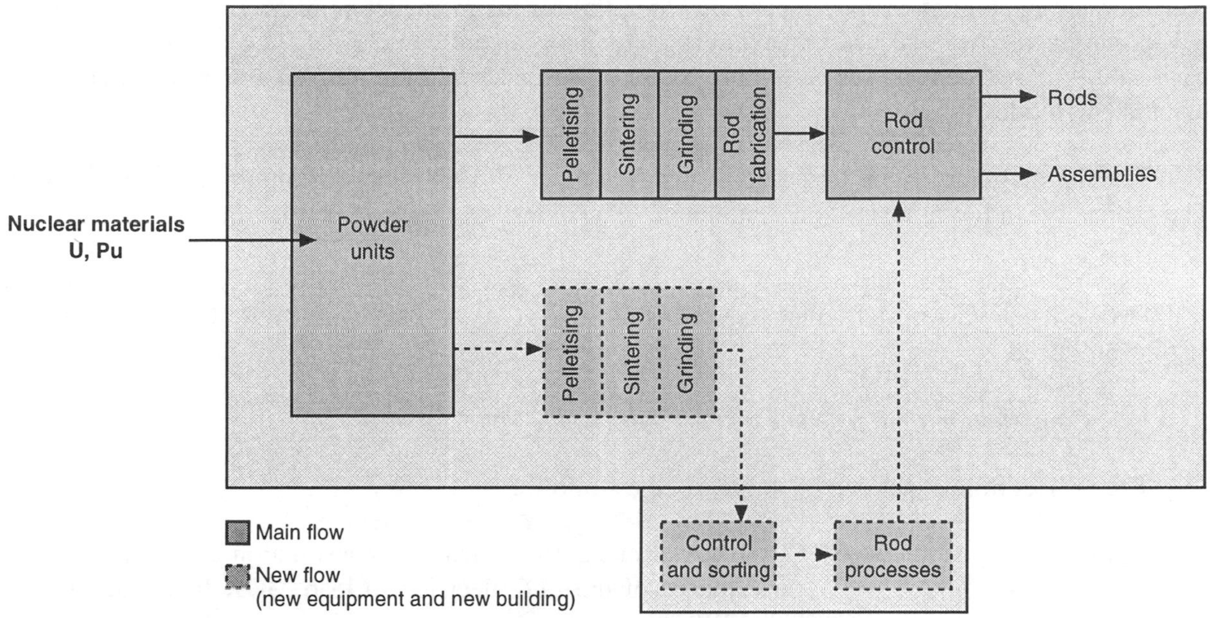

The objective is to recycle most of the scraps in the process line and recover as much plutonium as possible from the waste (five kg per year, *i.e.* one thousandth of the plutonium main stream), to reduce their volume and toxicity and compact them for final conditioning. Low-activity liquid waste will be treated in the COGEMA Marcoule treatment unit.

An auxiliary building on-site is devoted to waste and scraps conditioning, including a large incinerator with a capacity of 20 kg/h for the alpha-contaminated burnable waste. Concerning the scraps, when they are not directly recycled, they will be chemically treated at La Hague.

b) Design and safety bases

The reference starting materials for the production of MOX fuel in MELOX are the following:

- Rod hardware: lower end plug welded cladding tubes, upper end plugs, plenum springs (and other rod hardware, if any) are received from usual UO_2 fuel manufacturers.

- Assembly hardware: skeletons, top and bottom nozzles and small assembly hardwares are received, as well, from usual UO_2 fuel manufacturers.

- UO_2: originates preferentially from depleted uranium, tails from enrichment plants, but natural uranium or reprocessed uranium are also acceptable, provided that the ^{235}U content is less than 1 w/o in uranium. The UO_2 must be free flowing. Such free flowing UO_2 is, as reference, to be delivered by the TU2 conversion COGEMA Pierrelatte plant. The conversion process is of the ADU type conversion process.

51

- PuO_2: the PuO_2 is to be received in cartridge boxes from the COGEMA reprocessing plants in La Hague (UP2 – UP3).

Because of the new trends in fuel management and the need to offer utilities a flexible use of MOX fuel, the design of MELOX has been based on the use of a wide range of basic nuclear materials (for instance, uranium being either depleted from enrichment plants or from reprocessing plants, PuO_2 arising from the reprocessing of high burn-up UO_2 fuel, after storage of five to six years, with americium content of up to 3 per cent $^{241}Am/Pu$) for the manufacture of MOX fuel to be used in LWRs within the next 20 years (maximum plutonium content could reach 11 per cent on fuel rod basis and 9.75 per cent on fuel assembly basis). The maximum mass of plutonium allowed in the plant is 14 t oxide.

The thermal studies are based on a specific power of 19.9 W/kg maximum in the plutonium. The criticality studies are based on a minimum ^{240}Pu content of 17 w/o in plutonium.

As a reference, the outputs are finished MOX fuel assemblies for 17 x 17 type PWRs (900 or 1 300 MWe). However, MOX fuel assemblies of 16 x 16 or 18 x 18 type PWR (German type) can also be produced.

c)   *Construction time schedule and present status*

The project milestones can be summarised as follows:

1985:	COGEMA decided to launch the construction of MELOX as a consequence of the EDF decision to load MOX fuel in 16 PWRs (900 MWe).
1988:	Issue of the preliminary Safety Report together with detailed preliminary design for production, scrap and waste processing.
February 1989:	Issue of construction licence.
May 1990:	Issue of licence as a basic Nuclear Facility.
Beginning of 1991:	Start of construction of the production building and detailed preliminary design for the waste conditioning/incineration building.
September 1992:	Beginning of construction of the waste conditioning/incineration building.
June 1993:	Issue of provisional Safety Report.
November 1993:	Introduction of dummy UO_2 rods (for testing equipment for NDT of rods and assembling).
1st quarter 1994:	Introduction of depleted UO_2 powders in the process of testing equipment.
4th quarter 1994:	• Introduction of PuO_2 powder in the process and process qualification for the production of MOX fuel.

> ◆ Qualification of the MOX assembling process with the use of MOX rods delivered by the CFCa plant.

1st & 2nd quarters 1995:   Start-up of industrial production with the qualification of pellet and rod fabrication steps.

Mid-1995:   Decision to extend capacity of MELOX up to 160 t HM per year, to be in operation in 1999.

The plant is owned and operated by MELOX, a subsidiary of COGEMA (50 per cent) and FRAMATOME (50 per cent). The licences to construct and operate the plant have been granted to COGEMA.

2.5   Construction of new MOX fuel fabrication plants

As can be seen in Table 3, additional modern MOX fuel fabrication plants are expected to be in operation within the next ten years. Their anticipated characteristics and the experience gained with their design and construction are presented in this section. The following plants are reviewed: SIEMENS Hanau, SMP Sellafield, Belgonucléaire P1, the future Japanese plant and the COGEMA MELOX-extension.

Their conceptual design has the following common bases:

- The implemented fabrication process relies on processes and techniques already used and qualified in industrial scale plants: SIEMENS Hanau integrates the experience from the smaller MOX plant on the same site. MELOX and the MELOX-extension use the process and technology of Belgonucléaire P0 for MOX fuel and those of FBFC for fabrication of fuel rods and assembling; the SMP process and technology are based on those from MDF which is located on the same site; Belgonucléaire P1 implements the same process and technology as the P0 plant, but in an optimised layout as a result of P0 operation experience; and the Japanese future plant is to be based partially on PNC MOX technology and the Japanese UO_2 fuel manufacturers technology, while also relying on the expertise of European MOX fuel manufacturers.

- The enhancement of the protection against earthquake, aircraft crash, pressure wave (external explosion) and fire: the result is a more massive building, in particular when the fabrication process is distributed over several floors.

- The improvement of radiation protection to take into account:
  a) the use of the plutonium arising from the reprocessing of high burn-up spent UO_2 fuel;
  b) the manufacture of MOX fuel designed for increased discharge burn-up; and
  c) the more severe radiation protection recommendations (ICRP-60). The average radiation exposure of the group of workers associated with the fabrication plants should not exceed 5 mSv per year (whole body dose).

This constraint results in extensive process automation or remote control of the operations. Increased automation has the potential to increase dose uptake to maintenance staff. Detailed design is used to ensure that maintenance times and frequencies are minimal. Modular design for equipment, together with maximising access within the constraints of containment, are essential considerations in that respect.

The motivation and the specific context in which the decision to start the new MOX plant projects was taken and according to which the detailed design was carried out are significantly different from one case to another and explain somehow the large difference of the layout adopted for those plants:

- The SIEMENS Hanau plant was conceived essentially to recycle the plutonium from the German utilities (from La Hague or THORP reprocessing plants) in German PWRs (16 x 16 or 18 x 18 types).

- MELOX has been designed to recycle the plutonium from EDF (from La Hague reprocessing plant) in EDF reactors (17 x 17 types). An extension has been designed to recycle the plutonium in other reactor types (PWRs and BWRs).

- The SMP plant will manufacture MOX fuel with plutonium arising mainly from the THORP reprocessing plant, for use in a broad variety of foreign LWRs (but based essentially on PWR fuel experience).

- DESSEL (P0 and P1) is basically to offer its MOX manufacturing services to a broad variety of fuel vendors for use by a large variety of utilities in various countries. This requests a large flexibility in terms of fuel designs (PWRs of 14 x 14, 15 x 15, 16 x 16, 17 x 17, 18 x 18 types, and BWRs of 8 x 8, 9 x 9, 10 x 10 types) and in terms of fabrication campaign sizes.

SIEMENS – Hanau (30, 31, 32)

New fabrication facility

In 1982, construction of a new MOX fabrication plant was decided to meet the changing requirements with regard to throughput, fuel burn-up and safety issues. The main guideline was to adequately keep the philosophy and the basic proven techniques of the existing plant, while improving the economics, quality assurance, radiation protection, safety and security.

The building for the new fabrication plant has been constructed on the same site as the existing plant, next door to the existing plutonium storage building. The plant contains two production lines arranged in three levels to save floor space: production flow is from top to bottom.

The second floor accommodates filter housing linked to the adjacent supply systems building, a new laboratory wing for analysis during automated operation, and a small plutonium laboratory for research and development purposes. The uranium and plutonium reception area on this floor represents the start of the fabrication process for the OCOM method.

The first floor houses the equipment for powder preparation, pressing, sintering and grinding installed in separate rooms, in two parallel lines. The pellets are then transferred by a vertical transport system (designed as a bull head to withstand aircraft crash) to the ground floor where they are assembled in fuel rods.

Linked to the fabrication building is the plutonium store, completed in 1979 and comprising one fully bunkered building and a less heavily protected building. In the future, the fully bunkered building would have storage rooms for PuO_2, plutonium nitrate, fuel rods, fuel assemblies and returns. In addition, one of the rooms was to house a plutonium conversion plant operating on the AUPuC principle with an annual capacity of ten tonnes of plutonium/uranium mixture (plutonium content of 30 per cent). This plant is installed to convert the residual plutonium nitrate from the reprocessing plant at KfK and to return wet waste; it also includes a hydrofluoric acid dissolving station for material that is not fully soluble in nitric acid.

The non-bunkered section of the fissile material store contains the assembly room; in the future, the fuel assemblies were to be assembled vertically in a pit. The plutonium limit for this section of the plant is 99 kg.

The production facilities were to be removed from the caissons of the existing production facility (see section 2.4), so that the vacated caissons could be used for a uranium store, a cladding tube processing facility, and a disassembly cell.

Layout and design principles

The plant has been designed for a production capacity of 120 t HM per year of MOX fuel, typically for 16 x 16 PWRs (reference PWRs in Germany). PWR and BWR-MOX fuel can be produced with an average content of 5 w/o fissile plutonium, but also high burn-up fuel, with plutonium contents of about 10 per cent and burn-up higher than 50 GWd/t.

For criticality safety purposes, a plutonium vector of 95 per cent ^{239}Pu and 5 per cent ^{240}Pu is considered (as for the existing plant).

The calculation of personnel doses is based on the following isotopic composition (reference plutonium vector): ^{238}Pu to ^{242}Pu: 1.5 – 58.6 – 23.8 – 11.0 – 4.8 per cent and ^{241}Am: 0.3 per cent.

For the plant design the radiological relevant source strength was set at 125 per cent compared to this reference, and the americium value to 1.75 per cent (maximum permitted 3 per cent).

The maximum value of the personnel effective dose for normal operation was set at 10 mSv per year.

Present status

The first partial permit was granted in October 1987 and the construction started in December 1987. The five further partial permits were granted between April 1989 and March 1991. They permitted the construction and subsequent operation of the entire plant with the two new production lines, including the conversion area using the AUPuC process. A time schedule for the implementation of the permits had been agreed upon with the licensing authorities in order to reach full capacity sometime in 1997:

- initial start-up of the new line 1 with plutonium by the end of 1992;
- corresponding start-up of the new line 2 in spring 1993;
- reconstruction of the assembling room to accommodate vertical automated assembly in the first quarter of 1993 and start-up by the end of 1993; and

- shut-down of the old plant, after start-up of the second production line in the new fabrication building, followed by decommissioning.

However, plans changed with the election in the Hesse state in January 1991, after which the local government had been acting to have the production of MOX fuel elements abandoned in Hanau.

The authorities began what is called *"law enforcement towards abandoning of nuclear power"*. The result today is that 95 per cent of the new plant is completed, and that SIEMENS and the German utilities have decided not to operate it.

SMP – Sellafield MOX plant (38)

Fabrication facility

The source of plutonium dioxide will be, primarily, the THORP finishing line, but provision is also being included to process materials from reprocessing of Magnox fuel and also to receive oxide from other sites, if the customer so requires. The reference specification for the design safety case is the plutonium arising from UO_2 fuel irradiated up to 45 GWd/t burn-up with material subsequently aged five years (30 000 ppm ^{241}Am). In addition to this, the design will also cater for higher burn-up (55 GWd/t) and longer aged oxide (ten years, 45 000 ppm ^{241}Am), on a non-routine basis.

The source of uranium dioxide is material prepared by the integrated dry route process in Springfields. Depleted tails or natural enriched powders are anticipated to be the main feed, but the design will allow for the receipt of recycled material.

The product range for the plant includes both PWR output and BWR complete fuel assemblies and, in addition, the plant will have capability to produce pellets for other reactor types, including AGR and fast reactors. It will also be possible to manufacture fuel to advanced specifications containing up to 10 per cent fissile plutonium for LWRs. The reference production capacity is 120 t HM per year.

The flow-sheet for the Sellafield MOX plant is similar to that for the MDF. It incorporates the short binderless route, but includes a powder bulking-up stage to reduce analysis requirements. Provisions in the SMP consist of two parallel process lines with the important inclusion of a recycle capability for pellet material (rejected at inspection stages) and other "clean" arisings throughout the facility.

Overall waste arisings are relatively small, but a significant proportion of the solid plutonium contaminated arisings are the empty cans following plutonium oxide removal. These will be despatched automatically to the waste handling area of the plant, to minimise both additional waste and potential dose uptake. Bagless transfer facilities will also be incorporated to reduce waste arising for movement of items between glove boxes.

Layout

Plutonium dioxide cans are elevated vertically within THORP and transferred across through a duct into the SMP facility to the receipt box. Uranium dioxide and plutonium dioxide from elsewhere come directly to the SMP and are elevated within the building, subsequently being fed to opening and dispensing boxes. The gravity linking of the mill, blend, conditioning and pellet press stage is from the 18 m down to the 6 m level, with pellet handling largely confined to the 6 m level. Recycled material, which is primarily crushed pellets, is transferred back up to the 12 m level. Finished pellets are transferred downwards from the pellet tray store to the zero meter level where stack preparation, rod filling and inspection are carried out. Rods are then loaded into the magazine store from where they are taken to manufacture fuel assemblies. The fuel store is also included at the zero meter level, from where assemblies can be removed and packed for despatch.

Project status

- The project was approved in June 1993.

- Full planning permission has been granted, regulatory approvals have been obtained and construction has commenced in April 1994.

- Construction is to be complete by July 1996.

- Commissioning under uranium is planned starting May 1997.

- The plant is planned to start operation in late 1997.

Belgonucléaire – P1 (23, 24)

Fabrication facility

The development of additional MOX fuel production capacity to the present P0 plant is considered on the Belgonucléaire-Dessel site with the detailed design of the so called P1 extension. A connection is provided between the two plants, which allows for possible transfer of products between them.

Such extension is designed for a reference production capacity of MOX fuel of 40 t HM per year. The total production capacity of the P0 and P1 plants is to reach 75 t HM per year.

Basically, P1 is essentially implementing proven processes (and in particular the MIMAS process) and technology already in operation in P0. The plant is arranged in one level layout, with a fully flexible architecture organised along two manufacturing lines to have the maximum flexibility in operation.

The implementation of these processes is designed to take into account the more severe constraints to be met in the future and is thereby the continuity of the progressive evolution of the P0 plant during its 20 years operation.

These constraints are mainly:

- The use of plutonium arising from the reprocessing of increased discharge burn-up (*i.e.* 45 000 MWd/t) spent UO_2 fuel, leading to degraded isotopic composition and higher radioactivity and thermal specific power.

- The manufacture of MOX fuel designed for increasing discharge burn-up, leading to higher fissile plutonium contents in the MOX fuel (up to 10 w/o Pu+Am/U+Pu+Am).

- The more severe radiation protection regulations (ICRP-60).

The P1 plant design differs, however, from that of P0 in several specific aspects :

- The layout is optimised as a result of the lessons learnt from the operation of P0. Intermediate storage of the products in the fabrication line at several steps of the fabrication process allow the creation of buffers between the equipment. These buffers make it possible to cope with the different instantaneous throughputs of individual equipment and to pursue the production during limited maintenance of equipment.

- In P0, progressive automation or remote control of operations is being implemented. P1 will benefit from this evolution by automatising the most repetitive tasks, so that under normal conditions and after proving the reliability of the mechanisms, the equipment can be producing in nominal conditions.

- Machine coding of containers such as cans, sintering boats, pellet trays and cladding tubes will allow automatic identification of the products each time they leave or enter the equipment and thereby contribute to fully computerised data acquisition for product traceability and fissile material balance.

- Provision is made for further installation of advanced automated inspection, such as computerised visual pellet inspection and computerised end plug weld radiography, when these will have been proven fully operational in industrial production.

- A highly flexible and redundant transport line serves the various materials preparation halls and equipment. It allows the transfer backwards and forwards of the products between the various work stations and from and to the in-line storages. This transport line can be remotely controlled.

- Compared to P0, the safety is to be enhanced with regards to the protection against earthquake and aircraft crash.

Project status and construction time schedule

The P1 plant extension design is presently being finalised. Belgonucléaire, in cooperation with COGEMA, is defining the conditions of the construction of P1: operation to start at the turn of the century with an output of 40 t HM per year, leading to a total production of MOX fuel on the Dessel site of 75 t HM per year.

The planning schedule for the construction of the P1 plant is such that production on the first line can start four years after starting the project and about one year later on the second line. The production of fuel by the P1 plant will be 10 t HM in the first year after qualification of the first line, 25 t HM the year after, and will then reach its full capacity of 40 t HM per year.

The main characteristics of the Dessel plants (Belgonucléaire P0 and P1 for MOX fuel in four production lines, and FBFC for mounting assemblies of various designs for PWRs and BWRs) are their flexibility and their ability to handle various kinds of fuel designs and batch sizes.

The Japanese plant

The Japanese Atomic Energy Commission issued the *"Long-Term Programme for Research, Development and Utilisation of Nuclear Energy"* in June 1994. Regarding the construction of the commercial MOX fabrication plant in Japan, the programme notes that: *"considering the plans for utilisation of MOX fuel in light-water reactors and operation of the Rokkasho reprocessing plant, it will be necessary to establish commercial fabrication of MOX fuel in Japan on a scale of a little under 100 t per year by shortly after the year 2000".*

COGEMA – MELOX extension

The MELOX plant has been optimised to produce large quantities of 17 x 17 PWR fuel. It constitutes a substantial contribution to the experience gained in the construction of new MOX fabrication plants and is described in some detail in section 2.4.

In order to extend its services in the MOX field to any PWR design, as well as to BWR fuels, COGEMA has undertaken an extension of the MELOX plant, adding 50 t HM per year of capacity by the end of the century. The total plant capacity will reach over 200 t HM per year with this addition.

The MELOX extension consists of an additional new pelletising, sintering and grinding line within the existing building and in a new building including pellet controls and sorting, and fuel rod fabrication (Figure 4).

The extension will use the head-end and back-end stages of the fabrication process of the MELOX plant, *i.e.* the production of the primary blend and the secondary blend, on the one hand and the rod testing and assembling processes, on the other.

This multi-design concept would allow manufacturing of pellets in various diameters or lengths and producing fuel rods of different lengths, for example from two to four meters long. The extension, decided in May 1995, is to be operational in 1999.

Impact of the plutonium isotopic vector on the MOX plant design

In comparison to plutonium commonly known as reactor-grade, considered above, plutonium from low irradiated fuels is characterised by:

- A much larger proportion of the ^{239}Pu isotope, leading to a strengthening of criticality-related safety constraints.

- A significantly lower amount of ^{238}Pu, ^{240}Pu and ^{242}Pu isotopes, reducing the alpha activity and the neutron flux coming from both spontaneous fissions and alpha-n reactions.

- A significantly lower amount of ^{241}Pu, resulting in very low ^{241}Am build-up, and in a significant decrease of the gamma irradiation level.

- A lower Pu/HM content in the MOX to be fabricated for the same design basis.

The processing of plutonium separated from very low irradiated fuels would, therefore, lead to an alleviation of MOX plant design requirements and fabrication constraints derived from irradiation levels, namely, biological protection and cooling requirements.

Figure 5.   Build-up of plutonium isotopes in a UOX-FA of 4.0 w/o ^{235}U initial enrichment

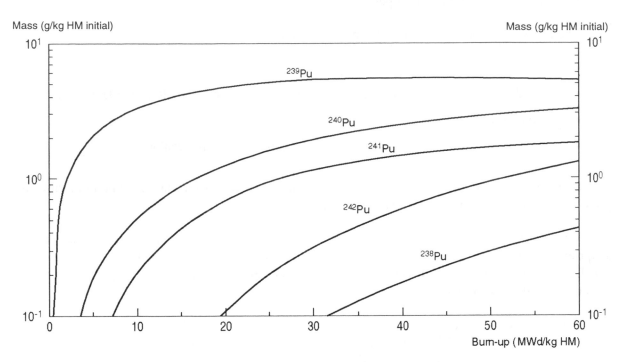

Dimensions, e.g. fuel rod diameter, pitch or fuel assembly structure, are more or less similar in current reactors, but can be adapted in new reactor systems designed to improve utilisation of MOX fuel.

The control and safety characteristics of the core are dependent on the proportion of MOX fuel in the core. The fuel assemblies must be designed in accordance with the whole core. Today, a 30 per cent MOX loading requires small modifications, in particular, in the number and the location of control rods. Beyond 30 per cent (depending on the reactor design) more important changes of safety related components become necessary.

Plutonium utilisation strategies

Since any design of MOX fuel assemblies and MOX containing cores has to achieve the same safety requirements as UOX cores, plutonium-bearing fuel rods and fuel assemblies have to meet the same thermal and mechanical limits as those specified for UOX fuel.

Specific additional costs associated with the fabrication of MOX fuel (which are believed to be significantly reduced at increased throughputs), give an incentive to select the plutonium concentrations as high as possible within nuclear and thermal limitations.

In relation to the MOX content of the cores, the following cases are of interest (39):

- Self generated recycling (SGR) assumes that only a plutonium quantity equivalent to the previously generated plutonium of the same power station will be recycled, after a delay of some years for reprocessing and fabrication.

- Open market recycling (OMR) allows the earlier start of recycling, even with the first reload, and the concentration of plutonium produced by several UOX-fuelled reactors into a reduced number of selected reactors, which are then fuelled with a higher proportion of MOX fuel than in the SGR scenario.

- A thermal plutonium burner with 100 per cent MOX fuel (without UOX fuel assemblies) is the limiting case of OMR and allows special designs of MOX fuel assemblies and cores to achieve the safety requirements (e.g. by appropriate MOX fuel assembly optimisation and upgraded control rod and boron systems, when needed).

MOX fuel assembly concepts for PWRs and BWRs

In order to make the use of plutonium in LWRs as competitive as possible, there is a need to concentrate the plutonium in a minimum number of fuel rods, as well as in a minimum number of fuel assemblies, to minimise the higher costs for fabrication and transportation.

The interface with UOX fuel rods and especially water gaps between assemblies and inner water areas in assemblies (as in BWRs) influences the MOX fuel assembly concepts.

Two assembly configurations have been investigated and tested in the past (21):

- The "plutonium island" assembly. This is an assembly with the MOX fuel rods located in the central zone and enriched uranium rods at the periphery. It seems to be more convenient for BWRs or any reactor having large water gaps between assemblies.

- The "all plutonium" assembly. This is an assembly comprising MOX fuel rods only. It is more appropriate for PWRs, where the effects of guide thimbles on flux and power peaking are corrected by an adequate choice of enrichment mapping. Designs for BWRs also use this configuration type.

In the central part of a MOX rod area, the plutonium content is selected high enough to reach the needed reactivity and burn-up. There is a tendency to have rather large areas of this kind. This favours fuel assemblies with MOX rods only, where the fissile plutonium content must be lowered only in the outer rods to avoid the power peaking induced by the thermal flux of the surrounding uranium fuel and moderating water areas.

In the past, a great fraction of MOX fuel assemblies have used natural uranium as the carrier material. Depleted uranium produced as tails from uranium enrichment plants, a waste product with a ^{235}U concentration of 0.2 to 0.3 per cent, is now considered more attractive as carrier material and reached today more than 60 per cent of all MOX production. Owing to their low fissile uranium content, uranium tails offer the opportunity of maximising the plutonium content.

There is, at present, a general tendency toward increased burn-up levels. This is especially advantageous for the economy of MOX fuel assemblies and leads again to higher fissile plutonium concentrations.

Figures 6 to 8 show examples of the structure of MOX fuel assemblies in use in Belgian (40), German (41, 42) and French (43) PWRs. Figure 9 gives the characteristics of MOX assemblies fabricated for insertion at Gundremmingen B/C in 1994 and Figure 10 gives the characteristics of MOX assemblies with improved water structure for future use in BWRs (41, 42). MOX fuel assemblies of SVEA design for improved inner moderation are also in the licensing process.

Figure 6. MOX-FA in use in Belgian 900 MWe PWRs

Type of rod		Number of rods
■	Zone 1: Low plutonium content	12
⊠	Zone 2: Medium plutonium content	68
□	Zone 3: High plutonium content	184
●	Guide tube	
Ⓘ	Instrumentation tube	

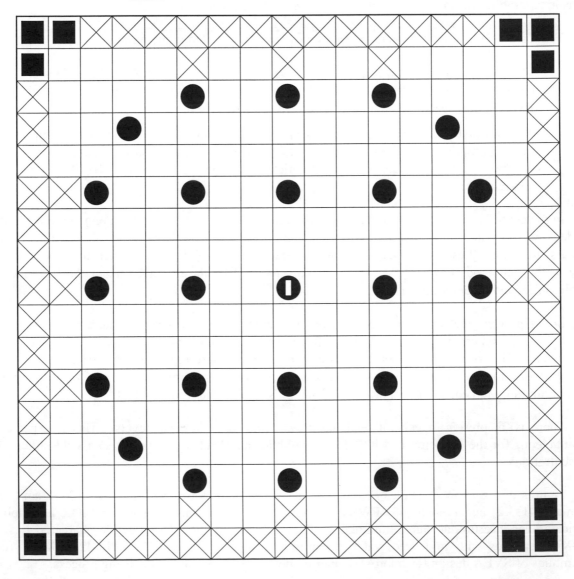

Figure 7. **MOX-FA in use in German 1 300 MWe PWRs**

Type of rod	Content of fissile material (w/o)		Number of rods
	Fissile plutonium (Puf/HM)	^{235}U ($^{235}U/U_{tot}$)	
■	2.0	0.25	12
⊠	2.8	0.25	92
□	4.1	0.25	128
○	–	Guide tube –	20
W	–	Water rod –	4

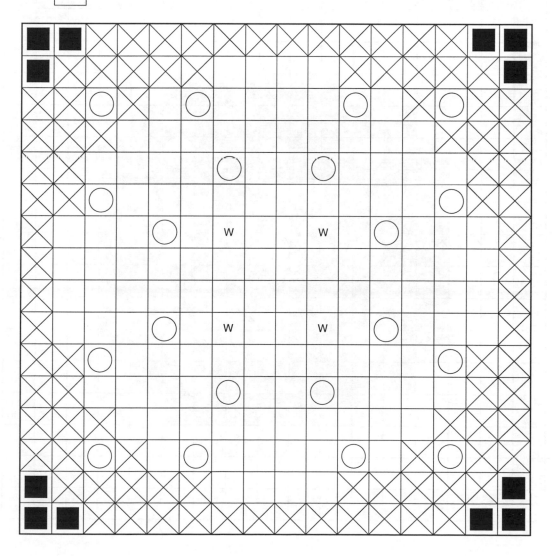

65

Figure 8.   **MOX-FA in use in French 900 MWe PWRs**

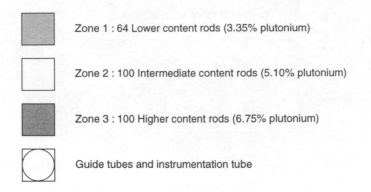

Zone 1 : 64 Lower content rods (3.35% plutonium)

Zone 2 : 100 Intermediate content rods (5.10% plutonium)

Zone 3 : 100 Higher content rods (6.75% plutonium)

Guide tubes and instrumentation tube

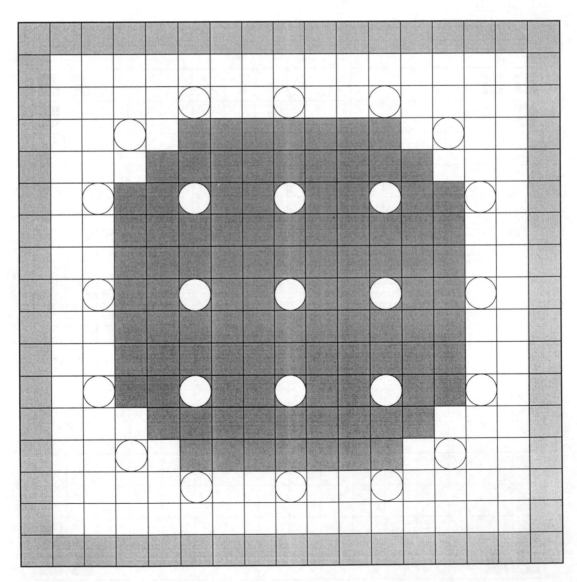

Figure 9. **Existing MOX-FA for Gundremmingen B/C BWR**

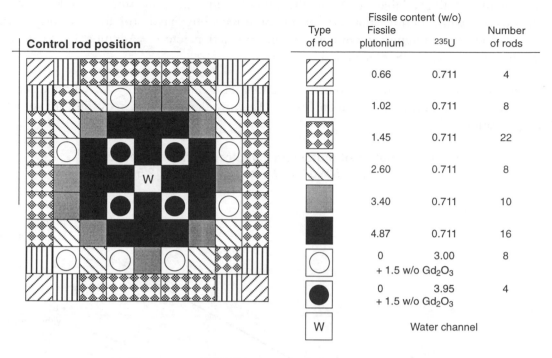

Type of rod	Fissile content (w/o) Fissile plutonium	^{235}U	Number of rods
(hatched)	0.66	0.711	4
(vertical lines)	1.02	0.711	8
(crosshatch)	1.45	0.711	22
(diagonal lines)	2.60	0.711	8
(grey)	3.40	0.711	10
(black)	4.87	0.711	16
(open circle)	0 + 1.5 w/o Gd_2O_3	3.00	8
(filled circle)	0 + 1.5 w/o Gd_2O_3	3.95	4
W		Water channel	

Figure 10. **MOX-FA proposed for German BWRs**

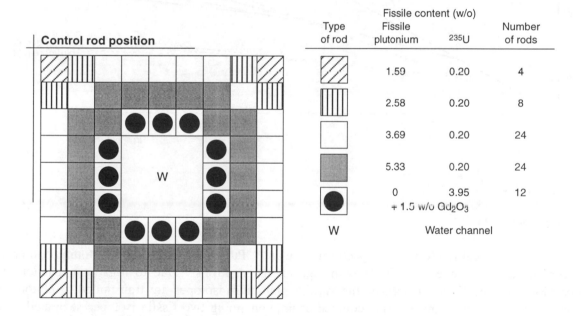

Type of rod	Fissile content (w/o) Fissile plutonium	^{235}U	Number of rods
(hatched)	1.59	0.20	4
(vertical lines)	2.58	0.20	8
(white)	3.69	0.20	24
(grey)	5.33	0.20	24
(filled circle)	0 + 1.5 w/o Gd_2O_3	3.95	12
W		Water channel	

Improved fuel utilisation, multi-recycling strategies

The present trend is to increase the fissile content of the fuel assemblies to reach higher burn-up. This trend is of even higher interest for the MOX fuel assemblies.

The concept of a closed fuel cycle with reprocessing and recycling implies also the reprocessing of MOX fuel assemblies (41). Separate reprocessing of MOX would lead to extracting plutonium of

67

poor quality (low fissile content, higher ^{238}Pu content) from the fuel cycle. As present practices foresee the mixture of all available plutonium, this would avoid at the same time a fast degradation of plutonium quality, while optimisation criteria of multi-recycling have still to be clearly defined (reactivity coefficients, limitation in minor actinide production, etc.). Additional studies of the nuclear related aspects would be needed in this context.

The resulting changes in the plutonium inventory of a PWR fuel cycle, without and with recycling as function of time, indicated in Figure 11, show the effect of recycling up to a third recycling generation.

Figure 11. Evolution of plutonium inventory with multi-recycling.
1 300 MWe PWR, 45 MWd/kg HM, recycling delay 12.5 years

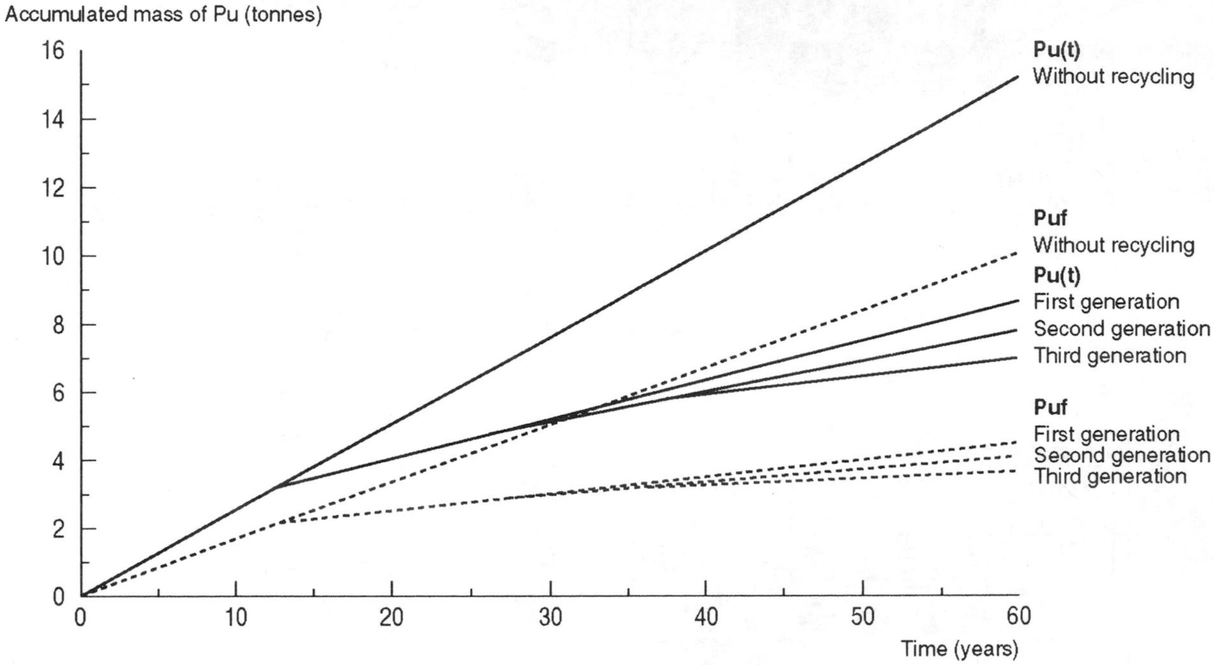

Neutron physics

General aspects (21)

Within the typical LWR neutron spectrum, ^{239}Pu and ^{241}Pu are the only fissile plutonium isotopes, although the other isotopes are fissile with high energy neutrons. Due to n-gamma reactions in competition with the fission reactions, the various plutonium isotopes are transmuted to plutonium isotopes of higher atomic mass. This coupled chain, containing two fissile isotopes separated by a fertile isotope, results in a variation of reactivity with burn-up which levels off much faster for MOX fuel than for UOX fuel. As a result, the reactivity of a core containing MOX fuel assemblies decreases less rapidly with burn-up than that of a core containing initially only UOX fuel, providing better stretch-out capabilities.

The variation of the cross-sections of plutonium isotopes with energy is more complex than for the uranium isotopes. The absorption cross-sections of the main isotopes (^{239}Pu, ^{241}Pu) are about twice as important as that for ^{235}U in the thermal energy spectrum, resulting in a relatively smaller reactivity worth for the control rods or for the boric acid in a UO_2-PuO_2 lattice. In addition, the absorption cross-sections of the plutonium isotopes are characterised by absorption resonances which are more numerous and much more important in the epithermal energy range (0.3 to 1.5 eV) than those of the uranium isotopes. Moderator temperature and fuel temperature (Doppler) coefficients are therefore more negative for MOX fuel than for UOX fuel. The overlapping of all the plutonium isotopes and of all their resonances makes the analysis of UO_2-PuO_2 lattices a challenge.

This section describes in some detail design issues and aspects of core management. Further details can be found in the OECD/NEA study on "Physics of Plutonium Recycling" (4) by the Working Party on the Physics of Plutonium Recycling of the NEA Nuclear Science Committee.

MOX design issues

The neutronic design of plutonium-bearing fuel rods, fuel assemblies and cores has to take into account the conditions provided by the established LWR technology. Changes in dimensions, e.g. of fuel rods, rod cells and fuel assemblies, could cause incompatibility to existing PWR and BWR systems and could only be realised in the case of new reactor systems.

The design of MOX fuel assemblies reflects the need of compatibility to UOX fuel assemblies, as long as MOX and UOX assemblies are in the same reactor at the same time. So, the MOX fuel assembly design is performed preferably in a geometric model including more than one assembly.

MOX fuel assemblies

The neutronic design of MOX fuel assemblies has to be adapted to the conditions imposed by neighbouring fuel assemblies (42, 43). As long as UOX assemblies are in the core, the transition of the UOX neutronic spectrum to the spectrum in the MOX fuel rod areas should be treated. The design and optimisation of the plutonium content, in the different types of MOX rods needed to avoid high power peaking in the areas of spectrum transition, is normally done by calculations with a sufficient number of (thermal) neutron groups in a geometry including UOX and MOX rods.

In the central part of a MOX rod area, the plutonium content has to be high enough for the needed reactivity (21, 42). There is a tendency to have rather large areas of this kind. This favours fuel assemblies with MOX rods only, where the fissile plutonium content must be lowered only in the outer rods to avoid the power peaking induced by the thermal flux of the surrounding uranium fuel.

As a consequence of the burn-up equivalence to be assured by core calculations, plutonium recycling does not affect the reload batch size (number of fresh fuel assemblies in a reload). A local power peaking form factor of 1.10 is obtained by using three plutonium content zones in the MOX fuel assemblies for PWRs.

The following two design aspects must be balanced:

- The power distribution within the MOX areas surrounded by uranium rods must be flattened out by using a minimum of different fissile plutonium contents and adjusting the distribution of the respective rods.

- The reactivity and burn-up potential of the MOX fuel assembly (FA) must be adjusted with respect to the uranium fuel assemblies via its average fissile plutonium content.

69

These aspects are valid for MOX-FAs for BWRs and PWRs, respectively (42). In BWRs the MOX-FAs are rather weakly influenced by the surrounding FAs, but strongly influenced by the water gaps between the FAs and void fraction inside FAs.

An extreme design situation is encountered, if a 100 per cent MOX core is to be designed. In this case there is no need of compatibility to UOX-FAs. Furthermore, the rod and cell geometry could be altered to optimise the core to plutonium use. As long as no other limits are reached, in this case one would use approximately the same moderation as with normal uranium feed LWRs. For PWRs, different plutonium contents in the MOX-FAs are not necessary, but a BWR plutonium-burner would need, as consequence of the water gaps, a plutonium variation over the MOX-FA, even in this case.

MOX containing LWR cores

Even for MOX-FA designs meeting the requirements of design and compatibility of UOX-FAs, the core properties are related to the core loading and to the fraction of MOX fuel in the core.

A prerequisite for plutonium recycling is the granting of a licence for the use of MOX fuel assemblies in the reactor, on the basis of given design requirements. Therefore, the technical feasibility is examined on the basis of realistic and enveloping designs. For this purpose, studies are carried out for different categories of requirements, as shown in Table 7 (42). On this basis, the validated limits of MOX fuel use are defined in the licensing procedure. Within these limits, licensing for individual cycles is then simply a matter of proving that cycle characteristics are within the limits analysed in the generic licensing evaluation.

Important cycle characteristics for various examples of PWR equilibrium cores with MOX fuel assemblies are listed in Table 8. The assessment of core characteristics is considered with reference to the differences between MOX and uranium equilibrium cycles. With MOX fuel assemblies in the core, the more negative moderator temperature coefficient and the smaller boron worth are especially apparent. As regards the net control assembly worth for the stuck rod configuration at EOC in the hot-standby condition, the data depend more on the loading scheme than on the fraction of MOX fuel assemblies in the core. Thus, the change from the traditional out-in reload pattern to a low-leakage one decreases the stuck rod worth enough to allow a MOX fraction of up to approximately 50 per cent without the need for more control rods.

Table 9 illustrates important characteristics of an equilibrium core with 31 per cent MOX fuel assemblies for a 1 300 MWe BWR. Core loadings have been investigated with up to approximately 50 per cent MOX fuel and fissile plutonium concentrations providing sufficient reactivity to be equivalent to uranium fuel designed for an average discharge burn-up of 45 MWd/kg. Especially in those cases of large amounts of fissile plutonium, it is important to mitigate the slightly less favourable neutronic characteristics of MOX fuel in comparison to uranium fuel, such as decreased control rod worth, burnable poison effectiveness and increased void reactivity feedback.

Transient and accident analyses for LWR cores containing MOX fuel showed only small differences compared with those for full uranium cores. For the rod drop accident as the limiting reactivity transient, the positive influence of the reduced control rod worth and the increased temperature and void reactivity feedback, as well as the reduced number of delayed neutrons are important. Altogether, no significant changes occur by the introduction of MOX fuel.

Under realistic operational conditions, the more negative void coefficient will even lead to a more favourable behaviour during that transient in comparison to a core without MOX fuel assemblies.

Table 7 **Safety evaluations related to MOX fuel assembly licences**

Areas of analysis	Categories of requirements		
	Normal operation		Accidents
	Reactor core	Spent fuel pool and new fuel store	Transients, LOCA, external events
Neutron physics	MOX-FA design Core characteristics	Sub-criticality Decay heat	Boron worth Reactivity coefficients Control rod worth
Thermal hydraulics	Unchanged	–	–
System dynamics	Control rod worth	–	as above
Fuel rod design	Fission gas pressure Corrosion	–	Fuel rod failure limit
FA structure design	Unchanged	–	Unchanged
LOCA analysis	–	–	Evaluated
Radiological aspectrs	Activity inventory	Activity inventory Release rates	Activity releases

Table 8 **Equilibrium fuel cycles for large PWRs with MOX**

Equilibrium fuel cycles in large PWRs with different MOX fuel assembly designs used in the licensing process

	48/25 Out-in-in	81/42 Low leakage with gadolinium	81/42 Low leakage with gadolinium	97/50 Part low leakage	193/100 Part low leakage	Effect compared with uranium cores
MOX fuel assembly loading (number/%)	48/25	81/42	81/42	97/50	193/100	
Loading scheme	Out-in-in	Low leakage with gadolinium	Low leakage with gadolinium	Part low leakage	Part low leakage	
Reload MOX and uranium fuel assemblies	16/48	24/32	24/32	24/24	64/0	
MOX fuel assembly type	16x16	16x16	16x16	18x18	18x18	
Fissile plutonium content (w/o)	1.9/2.3/3.3	1.9/2.3/3.3	2.2/3.0/4.6	2.0/2.6/ 3.9/5.0	4.1	
^{235}U content (w/o)	0.7	0.7	0.25	0.7	0.7	
Uranium fuel assembly ^{235}U enrichment (w/o)	3.4	3.5	3.5	4.0	–	
Cycle length (days)	329	310	318	323	454	Same
MOX fuel assembly burn-up						Burn-up about the same
MOX batch (MWd/kg)	37.4	35.3	37.3	48.2	49.8	
Maximum MOX fuel assembly (MWd/kg)	39.0	41.9	43.7	54.6	57.6	
Initial boron concentration (ppm)	1 247	1 088	1 085	1 256	1 996	Lower
Reciprocal boron worth, BOC (ppm/% $\Delta\rho$)	-135	-147	-158	-178	-298	Higher (about -120)
MTC at EOC (pcm/K)	-59.5	-69.1	-61.4	-77.4	-78.5	Higher (about -55 to -65)
Net control rod worth at EOC (% $\Delta\rho$)	5.5	6.6	4.7	5.4	5.3	Lower to same

Table 9 **Equilibrium fuel cycles for a large BWR with and without MOX**

Data of equilibrium fuel cycles in a large BWR with and without MOX fuel assemblies		
MOX fuel asembly loading (number/%)	264/31	0/0
Loading scheme	Low leakage	Low leakage
Reload		
MOX fuel assembly	40	0
Average fissile plutonium content (w/o)	3.26	–
Average ^{235}U content (w/o)	0.80	–
Uranium fuel assembly	96	136
Average ^{235}U enrichment (w/o)	3.4	3.4
Cycle length, including coastdown (days)	296	298
MOX fuel assembly burn-up		
MOX batch (MWd/kg)	45.2	–
Maximum MOX fuel assembly (MWd/kg)	47.3	–
BOC hot excess reactivity (% $\Delta\rho$)	1.3	1.1
BOC cold shutdown margin (% $\Delta\rho$)	1.4	1.3
MCPR		
Uranium fuel assembly	1.35	1.38
MOX fuel assembly	1.50	–
Maximum linear heat rate		
Uranium fuel asembly (W/cm)	412	437
MOX fuel assembly (W/cm)	399	–

Experience with MOX containing LWR cores

In Germany, after the early plutonium recycling programmes in BWRs (Kahl, KRB-A), the beginning of commercial MOX use in LWRs was concentrated in PWRs. Figure 12 gives a survey of the German programme (by SIEMENS), including the starting phase at KWO (NPP Obrigheim) and the Swiss plant Beznau-2 for the time interval 1972 to 1993. The three KONVOI PWRs are licensed in Germany for the irradiation of up to 50 per cent MOX-FAs of the type 18 x 18 and several BWRs have requested a MOX insertion licence (42) or have already received such a licence, e.g. KRB-B.

In parallel, four test MOX-FAs had been inserted (by Westinghouse) in 1978 in Beznau-1 and a further 60 MOX-FAs (by Belgonucléaire) had been loaded in this reactor in the time span of 1988 to 1992.

The French programme regarding the 900 MWe PWRs of EDF started with a first reload including 16 MOX-FAs in 1987 at St.-Laurent-B1 and comprises, at present, nine plants (see Figure 13) in which a total of 500 FAs (130 000 fuel rods) have been loaded. This programme included the insertion of a further 16 MOX-FAs in the SENA reactor during 1987 to 1991. Seven additional 900 MWe PWRs of EDF are licensed to recycle plutonium. This is depending essentially on the availability of MOX fabrication capacity.

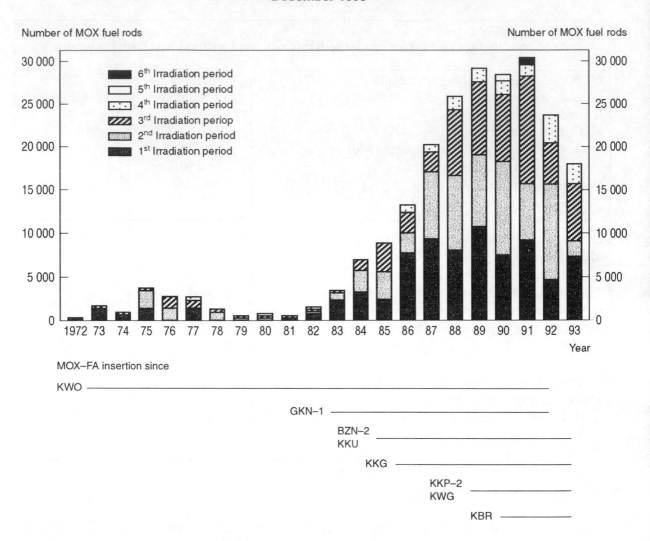

Figure 12. **MOX insertion in PWRs by SIEMENS KWU until December 1993**

In Belgium, after the 1963-1987 data base acquisition programme in the BR3 PWR (with up to 70 per cent MOX fuel in a reload), two 900 MWe PWRs have been loaded with MOX fuel since 1995 (40).

In Japan, MOX fuel has been loaded in the Tsuruga-1 (BWR) and the Mihama-1 (PWR) reactors which are both commercially operated. In Tsuruga-1 two test assemblies (8 x 8) and in Mihama-1 four assemblies (14 x 14) were irradiated for cycles in the years 1986 to 1990 and 1988 to 1991, respectively.

The neutron physics experience acquired in these reactors is based on:

•   start-up measurements;

•   in-service cycle monitoring; and

•   specific measurements.

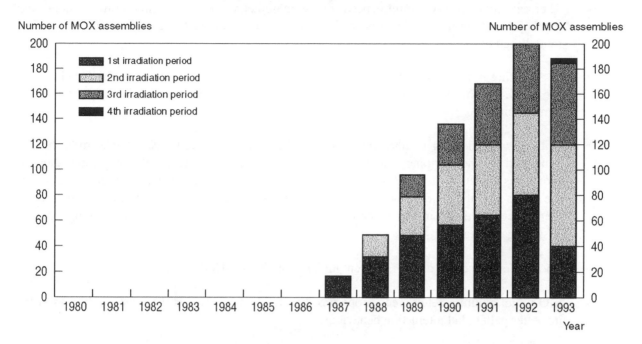

Figure 13. **Plutonium recycling in EDF 900 MWe PWRs –
Distribution of MOX reloads**

The reliability of the design methods is confirmed by measurements of cycle length, boron concentrations, reactivity coefficients (such as for coolant temperature and boron worth), control rod worth, and power density distribution.

No significant increase in the deviations between measurements and calculations were found with increasing the MOX content of succeeding cycles when modern data sets and calculation tools were used.   Examples are the French measurements on critical boron concentration, isothermal coefficient, and several control bank worth values (43, 44)..

For the validation of the power distribution (42, 43) two examples are given in Figures 14 and 15. They cover the cycle five of the German NPP Grohnde in full low-leakage loading and the EDF power plant St.-Laurent-B1 containing 3 x 16 MOX-FAs (cycle seven).

The task of validation is ongoing, as new MOX-FA designs, higher plutonium contents and higher percentages of MOX-FAs in the cores are reached.

Mechanical and chemical properties of MOX fuel and MOX fuel rods

General aspects (21)

Significant efforts have been devoted to determine the physical properties of MOX fuels for fast reactors which generally have a 20 per cent plutonium content. Studies have shown that several of these properties are poorer than those of UOX. Such data are, however, overly conservative for application to MOX fuel for thermal reactors which has a plutonium content of 4 to 10 per cent.

For such fuel, the performance and safety-related characteristics can be summarised as follows:

- Thermal conductivity is influenced by stoichiometry, porosity, plutonium content and irradiation. Correlations have been developed to incorporate the correct input in the fuel modelling codes.

- The melting point of stoichiometric MOX (at 5 per cent plutonium) is about 20°C below that for UOX. Within practical limits, hypostoichiometry does not have an effect on the melting temperature.

- Nuclear self-shielding is more pronounced in MOX fuel than in UOX. Hence, more heat is generated at the periphery in MOX fuels, mitigating the effects of the poorer thermal properties of MOX fuel at BOL. But, at EOL, the relatively lower depletion at the centre of the pellet results in more heat being generated at the centre and, combined to the lower thermal conductivity, in higher central and average fuel temperatures, therefore in higher FGR (fission gas release).

- Thermal expansion is about 1 per cent higher than that for UOX.

- MOX fuel exhibits better creep properties than UOX fuel. This is the most likely reason for its better pellet clad interaction behaviour.

- The homogeneity of the plutonium distribution in the uranium matrix of the pellets depends on the fabrication route. It affects directly fission gas release, densification and swelling, thermal limitation on particle size under reactivity excursions and the capability of the fuel to be reprocessed.

Experience has shown that a properly founded design technology, combined with adequate manufacturing techniques, can meet engineering and licensing standards of MOX fuel.

Chemical properties of plutonium are only relevant at the reprocessing step of the fuel cycle, where the solubility of the oxide and the treatment of plutonium nitrate and his conversion to plutonium oxide is needed.

Experience with in-core-behaviour of MOX fuel

The design of MOX fuel rods follows the procedure for UOX mechanical fuel and rod design. The small shifts in thermo-mechanical properties caused by the PuO_2 content, as mentioned above, have been accounted for. The less homogeneous structure of the fuel by master mix particles embedded in the UO_2 matrix is of no relevance, as long as the MOX particles are small enough to avoid local hot spots in the inner part of the pellets, as well as in its surface. This structure governs in particular the fission gas release, but also the dimensional behaviour during irradiation. As the power histories of MOX fuel rods tend, via less steep burn-down of reactivity, to higher power at higher burn-up, compared to UOX fuel rods, somewhat higher fission gas releases are calculated. The rod design has to take care of this effect by proper design of the plenum and fill gas pressure in the rods. The models used for the rod design have to be qualified by appropriate experimental data.

Figure 14. Validation of power distribution for MOX loading at Grohnde (cycle 5)

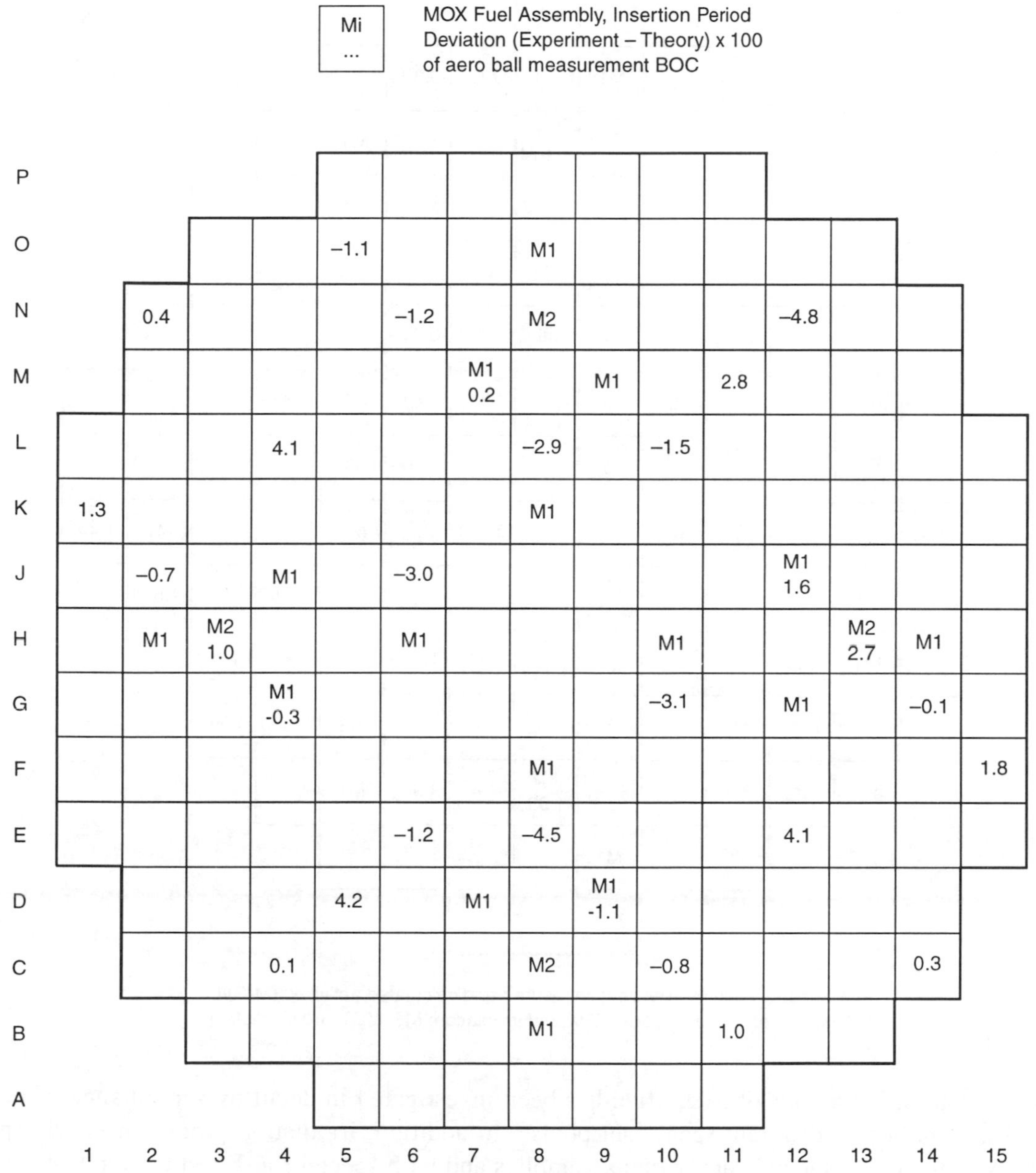

Mi ...	MOX Fuel Assembly, Insertion Period Deviation (Experiment – Theory) x 100 of aero ball measurement BOC

	1	2	3	4	5	6	7	8	9	10	11	12	13	14	15
P															
O				−1.1			M1								
N		0.4				−1.2	M2				−4.8				
M						M1 0.2		M1		2.8					
L				4.1			−2.9	−1.5							
K	1.3						M1								
J		−0.7	M1		−3.0					M1 1.6					
H		M1	M2 1.0			M1			M1		M2 2.7	M1			
G			M1 -0.3					−3.1	M1		−0.1				
F						M1						1.8			
E					−1.2	−4.5			4.1						
D				4.2	M1		M1 -1.1								
C			0.1			M2		−0.8			0.3				
B						M1		1.0							
A															

MOX loading pattern at KWG cycle 5 (20 MOX fuel assemblies):
validation of power distribution at BOC

	R	P	N	M	L	K	J	H	G	F	E	D	C	B	A
1								0.000							
2							M1		M1	-0.304					
3			0.008	M1			1.976	M2 -0.014			M1	-0.119			
4					M3 -4.753		2.833		0.842	M3					
5			M1 -1.228	M3	-1.136		M3 -3.300	M2	M3		-0.796	M3 -4.729	M1	-1.053	
6					-0.719		M2	2.924	M2	-0.416					
7		M1	1.081		M3	M2	2.865		3.346	M2	M3	-0.479		M1 -3.035	
8	-1.613		M2 -1.458		M2 1.070					2.209	M2		M2 -2.511	-2.691	
9		M1			M3 -3.382	M2			2.611	M2 -0.543	M3		M1		-1.239
10			1.147				M2 0.216		M2			-0.003		-1.675	
11			M1	M3	-0.087		M3	M2 1.157	M3	1.019	0.190	M3	M1		
12			-1.576		M3		0.873				M3	0.309	0.393		
13					M1			M2 -1.935		3.010	M1				
14					-2.831		M1	M1 -1.452							
15							-2.045								

Average deviation (%) between measured and calculated activities (M-C)/C
(M1: MOX 1st irradiation; M2: MOX 2nd irradiation; M3: MOX 3rd irradiation)

The irradiation behaviour of MOX fuel has been investigated in detail by surveillance of many MOX fuel assemblies in different spent fuel pools. In addition, irradiation programmes with path finder MOX fuel rods in special carrier fuel assemblies and with special MOX test rods in test rigs in selected nuclear power plants, as well as in test reactors, were performed followed by regular post-irradiation examination (PIE) in the spent fuel pool and in hot cells (40, 45, 46).

Investigations of MOX fuel rods show that the overall rod dimensional behaviour is similar to that of UOX rods. This is because identical cladding tubes were used for the rods examined, and the MOX fuel density behaviour was found to be roughly similar to UOX fuel up to burn-ups of

50 MWd/kg.   Understanding the density behaviour of MOX fuel requires consideration of its structure. On a microscopic scale, the fuel structure appears heterogeneous with MOX agglomerates uniformly distributed. Contribution of the high local burn-up in the MOX (master mix) particles, the matrix swelling rate related to the matrix burn-up, and the development of the porosity, sum up to the global dimensional behaviour of the MOX fuel.

Density measurements show a similar dimensional characteristic of OCOM (as well as comparable MIMAS MOX fuels) and AUPuC types of fuel. This results from the superposition of two-dimensional processes: the delayed densification and swelling of the matrix due to the low burn-up and the swelling of the MOX agglomerates.

An analysis of the fission gas release shows a strong influence of temperature. The highest gas release always occurred at the highest temperatures. Under steady-state conditions, this was concluded from the power history of the MOX fuel rods. Transient test conditions are better suited to quantitatively investigate the effects. MOX and UOX fuels showed similar fission gas releases at similar temperatures. This can be understood from the fact that the fission gas is always released via the UO_2 matrix. The remaining gas in the bubbles of the MOX agglomerates seems to be concealed, even at intermediate temperatures where the agglomerates are characterised by large bubbles within the agglomerates. Only at high temperatures, when release channels in the UO_2 matrix are formed, is the fission gas able to leave the fuel via these channels.

Transient tested MOX fuel shows a dimensional behaviour comparable to that of UOX fuel. The transient fission gas release was also found to be similar to that in UOX fuel operated at the same temperature.

The tight enclosure of MOX agglomerates by the UO_2 matrix, and the implantation of fission products and fission gases into the UO_2 matrix also prevent the instantaneous release of the fission products into the primary coolant in case of fuel rod defects and are considered to be the reason for the similarity in defect behaviour of UOX and MOX fuels.

In conclusion, a comparison of MOX and UOX fuels shows that both types of fuel, in spite of different structure and, hence, local burn-up, have similar dimensional and fission gas release behaviour. Therefore, it is justified from a technical point of view to also use similar models for design calculations.

Ongoing testing of MOX fuel is needed for the following reasons:

- small changes in pellet density and diameter, as well as canning dimensions and material properties, influence the burn-up behaviour;

- planned changes in the MOX fabrication technology need to be investigated;

- higher plutonium contents are to be included, at actual plutonium compositions; and

- an increase in burn-up is under preparation.

Summary

The experience, accumulated up to now, with MOX recycling which was presented above and the different strategies aimed at increasing the use of plutonium in MOX fuel for PWRs and BWRs, lead to the conclusion that MOX fuel is an industrial product, like uranium fuel. However, as is normal for every industrial product, future possibilities are discernible for improving MOX fuel in order to gain more efficiency and reactivity.

For countries and utilities involved in fuel recycling, the following main trends could be observed:

- Continuation with MOX recycling on a broad industrial basis in Europe and Japan. Currently, 32 LWRs, out of the world's total LWR number of 343, are licensed to use MOX fuel. Beyond the year 2000, it is expected that additional reactors would use MOX fuel in these countries.

- Improvement of fuel utilisation by multi-recycling strategies.

- Improvement of increasing burn-ups by using new MOX-FA designs with higher plutonium contents and higher percentage of MOX assemblies in the cores, up to 100 per cent.

- Stronger efforts towards standardization in order to improve economics.

Fast Reactors

Experience gained

The technology applied in Fast Reactors (FR) is familiar, as it has been in use for many years. The first FR was Clementine at Los Alamos (USA) in 1946 with a power of 150 kW. The first nuclear reactor to generate electricity in the world was a FR, the EBR-1 in the United States, in 1951.

From the outset, many types of fuel were tested: enriched uranium or plutonium in metallic, nitride, oxide or carbide form, or a mixture of plutonium and uranium oxides. Today, most FRs use the mixture called MOX (Mixed Oxide fuel), $(U-Pu)O_2$.

History of the FR

The FR history is as old as that of thermal reactors. For the first 20 years of their existence, these two systems advanced side by side. Around the sixties, as shown in Figure 16, four test fast reactors of about the same power size went critical: DFR in the UK (1959), EBR-2 in the USA (1963), Rapsodie in France (1967) and BOR-60 in the USSR (1969). The oldest and largest of them, DFR (72 MWt), was successfully operated over 18 years. So was also Rapsodie later on. BOR-60 is still in operation now. After DFR, which used sodium-potassium, sodium was retained as primary coolant.

The first prototype fast reactor for power generation was the US Enrico Fermi reactor (1964, 66 MWe). It had to be stopped in 1972 after an incident involving severe core damage. From 1972 to 1974, three prototypes of comparable size were successively brought into operation: BN-350 in the USSR (now in Kazakhstan), Phénix in France and PFR in the UK. The two first ones are still in operation now. The British prototype PFR was definitively shut down in March 1994, after 20 years of satisfactory operation. The SNR-300 prototype was built in Germany by a German-Belgian-Dutch consortium; plant and fuel were ready in 1985, but due to a political blockage the plant was never allowed to be started-up. The first criticality of the Japanese prototype Monju occurred in April 1994. Monju experienced a sodium leakage in the secondary loop in December 1995. Table 10 lists the main features of all fast reactors constructed in the world.

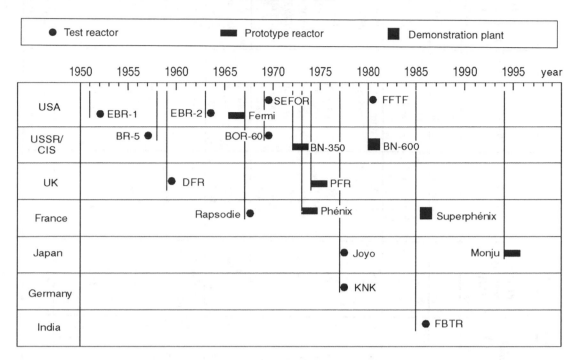

Figure 16. **Fast reactor programmes: start-up of reactors (first criticality)**

The third stage of development, the construction of a large demonstration plant, has been reached later on, in 1980, in the USSR (Russia) with BN-600 (600 MWe) and in 1985 in France with Superphénix (1 240 MWe). These achievements are further discussed below, together with those in Japan and other countries.

Obviously, the number of only two industrial-scale fast reactors is very small compared to the number of thermal reactors in operation in the world. On the one hand, these reactors give proof of the technical validity of the concept. On the other hand, the slowing down rate of construction of fast reactors has been associated with the need of mastering the technical difficulties of using sodium as a coolant, and with the relatively large investment costs of FR plants. The motivation for building fast reactors has also progressively changed.

Table 11 summarises the main features of fast reactors currently in operation.

The rationale for the FR

From the outset, the main objective for developing the FR was breeding. In this reactor, by converting ^{238}U (fertile material) to fissile ^{239}Pu, it is possible to produce more fissile fuel than is consumed. In slow neutron nuclear reactors of the PWR (Pressurised-Water Reactor) type, about 1 per cent of the natural uranium extracted from the mines is used to generate energy. The concern for a better management of uranium resources has first favoured the development of the FR, because with this reactor, the proportion of uranium actually used for power generation can increase, thanks to successive recyclings, up to about 60 per cent. It is easy to ascertain the advantage of such a technology in a world context of uranium shortage accompanied by large price increases, as was the forecast of nuclear energy needs in the seventies. It was therefore in anticipation of such an economic and energy environment that France decided, together with other European partners, to build the Superphénix industrial-scale reactor in 1977.

81

Table 10 **Main features of constructed fast reactors**

Reactor name	Country	Location	First criticality date	Shutdown date	Thermal capacity (MW)	Electric capacity (MW)	Fuel	Primary coolant	Primary coolant temperature (°C) In/Out
Clementine	USA	Los Alamos	1946	1953	0.025		Pu metal	mercury	140/40
BR-2	CIS	Obninsk	1956	1957	0.1		Pu metal	mercury	70/40
EBR-1	USA	Arco (Idaho)	1951	1963	1.4	0.2	U	sodium/potassium	450/375
BR-5	CIS	Obninsk	1959	1971	5		PuO_2, UC	sodium	
-10			1971		10		MOX, UN		
DFR	UK	Dounreay	1959	1977	72	15	U-Mo	sodium/potassium	350/230
EBR-2	USA	Arco (Idaho)	1963	1994	62	20	U-Zr, U-Pu-Zr	sodium	482/371
E. Fermi (EFFBR)	USA	Detroit	1963	1972	200	66	U-Mo	sodium	427/268
Rapsodie	France	Cadarache	1966	1982	20/40		MOX	sodium	510/404
BOR-60	CIS	Dimitrovgrad	1969		60	12	MOX	sodium	550/360
Joyo	Japan	Oarai	1977 (mark-I)		100 (Mark-II)		MOX	sodium	500/370
FBTR	India	Kalpakkam	1985		40		(U, Pu)C	sodium	518/400
KNK-II	Germany	Karlsruhe	1977	1991	58	21	MOX/UO_2	sodium	
SEFOR	USA	Arkansas	1969	1972	20		MOX	sodium	430/370
FFTF	USA	Hanford	1980	1994	400		MOX	sodium	590/370
PEC	Italy	Brasimone	Abandoned		125		MOX	sodium	525/375
BN-350	CIS	Chevchenko	1972		1 000	150 and desalination	UO_2	sodium	500/300
PFR	UK	Dounreay	1974	1994	600	270	MOX	sodium	560/400
Phénix	France	Marcoule	1973		560	250	MOX	sodium	552/385
SNR-300	Germany	Kalkar	Abandoned in 1991		770	327	MOX	sodium	560/380
BN-600	CIS	Beloyarsk	1980		1 470	600	UO_2	sodium	550/550
CRBR	USA	Clinch River	Abandoned in 1983		975	380	MOX	sodium	
Monju	Japan	Tsuruga	1994		714	280	MOX	sodium	529/397
Superphénix	France	Creys-Malville	1985		3 000	1 240	MOX	sodium	545/395
BN-800	CIS	Beloyarsk	Suspended			800	MOX	sodium	550/350

Table 11 Main features of fast reactors which are currently in operation

	Joyo (Mark-II) Japan	Phénix France	Monju Japan	BN-350 (Kazakhstan)	BN-600 (Russia)	Superphénix France
CAPACITIES						
Thermal capacity (MW)	100	560	714	1 000*	1 470	3 000
Gross electric capacity (MW)	0	250	280	150	600	1 240
Net electric capacity (MW)	0	233	246	135	560	1 200
CORE						
Active height/active diameter (m)	0.55/0.72	0.85/1.39	0.93/1.8	1.06/1.5	1.02/2.05	1/3.66
Fuel mass (tML)	0.76	4.3	5.7	1.17 ^{235}U	12.1 (UO$_2$)	31.5
Number of assemblies	67	103	198	226	370	364
Maximum power (kW/l)	544	646	480		705	480
Average power (kW/l)	475	406	275	400	413	280
Expected burnup (MWd/t)	75 000	100 000	80 000	100 000	100 000	70 000 (first core)
FUEL						
Fissile material	MOX	MOX	MOX	UO$_2$	UO$_2$	MOX
Enrichment (%) first core	30Pu	19.3Pu	15/20Puf			15.6Pu
Mass of plutonium (t) first core						6
Enrichment (%) reloads	30Pu	27.1Pu	16/21Puf	17/21/26	17/21/26	20Pu
Mass of plutonium (t) reloads						7
Assembly renewal rate	70 days	3 months	20% of core every 5 months	80 efpd	160 efpd	100% of core every 3 years
Form	pellet	pellet	pellet	pellet	pellet	pellet
Number of pins per assembly	127	217	169	127	127	271
Assembly geometry	Hexagonal	Hexagonal	Hexagonal	Hexagonal	Hexagonal	Hexagonal
Average linear power (kW/m)	40	45	21		36	
Maximum linear power (kW/m)			36	48	48	48
Maximum clad temperature (°C)	650	700	675	680	700	620
Maximum temperature at center (°C)	2 500	2 300	2 350	1 840	2 200	

* Real thermal capacity is 520 MW

In the eighties, the risk of uranium shortage was re-evaluated and appeared to be postponed by several decades, mainly because the growth rate of nuclear energy was lower in reality than that expected before. Uranium became abundant and cheap. Only part of the spent oxide fuel was reprocessed. The use of FRs in the "burner" mode gained in importance. While the presence of fertile blankets with depleted uranium around the core leads to plutonium breeding, the removal of these blankets can, without penalising core operation, reverse the plutonium balance: the reactor actually burns more fissile fuel than it produces. The fast reactor can thus be used to reduce the existing plutonium stockpiles.

Some facts about fast reactors

Unlike thermal reactors, FRs do not have a moderator. The lower fission probability is offset by increasing the fuel density in the core. In this type of reactor, all the plutonium isotopes are fissile to a varying extent.

Given the high power density, molten metals selected for their high thermal conductivity are used as coolant. The adoption of this type of fluid also helps to operate at low pressure, thus simplifying reactor vessel design. The first fluid used was mercury, but for technical and economic reasons, the metal used today is sodium (Na).

Sodium is a common and cheap metal. It melts at 98°C and boils at 880°C. Hence, it has a wide service range. Its density at these temperatures is comparable to that of water, so that all special pumping problems are eliminated. Yet sodium presents certain drawbacks: at the operating temperature, it ignites spontaneously in contact with air; also, the contact reaction with water is violent, and must be accounted for in designing the steam generators.

Today, the technologies associated with the use of liquid sodium are under control. The sodium that cools the core is activated and this entails the construction of thick shielding walls of concrete. The primary circuit is followed by a secondary inactive sodium circuit, which receives the power collected by the primary circuit through heat exchangers. The secondary circuits feed the steam generators. The primary circuit may be of the loop type (e.g. Monju) or built into the reactor vessel (e.g. Superphénix). The latter concept allows confining all the active sodium in the reactor vessel.

An advantage of the FR is its higher thermal efficiency with respect to the conventional reactors. The use of sodium at high temperature helps producing steam at higher temperatures and pressures. The thermal efficiency is accordingly about 40 per cent, versus about 33 per cent for LWR plants. Because of this, thermal emissions into the environment are lowered by 10 per cent.

Fuel

Owing to their respective nuclear characteristics, it is more advantageous, in energy terms, to burn the plutonium in fast reactors and the uranium in thermal reactors. Thus, in most fast reactors in operation today, the fissile nuclear material is plutonium, essentially recovered from thermal reactor spent fuel by reprocessing.

Because of fuel swelling at high burn-up and for compatibility with the clads, pure metallic Pu/U alloy is no longer used (see the IFR development). Today, fuels are essentially composed of ceramics obtained by sintering. The fuel most widely used at present is in the form of a mixture of plutonium and uranium oxides.

The fertile material is natural or depleted uranium oxide. It is found in the matrix of the fuel itself, and, for a breeder, in the blankets surrounding the core, in the axial and/or the radial directions.

The fuel assembly consists of a cluster of pins. Since the clad of the pins acts as a first barrier, it must be compatible with the fuel and the coolant (molten sodium) to guarantee mechanical strength and tightness for as long as possible. Stainless steel is the material that best meets these requirements today. As demonstrated by various tests, this material helps to reach burn-ups of between 100 000 and 200 000 MWd/t.

The pin is filled with helium and then closed at each end by a welded plug. Each pin is surrounded by a spacer wire or held in place by a stainless steel grid. A plenum in the pin serves as a reservoir for the released gaseous fission products and to maintain the internal pressure below the maximum permissible limit at the end of irradiation.

The pin cluster (127 for BN-350, BN-600, Joyo; 169 for Monju; 271 for Superphénix; 325 for PFR) is housed in a hexagonal wrapper tube that is also made of stainless steel. At the bottom, the assembly has a hollow foot that fits into the diagrid of the reactor, and maintains the assembly in the vertical position while allowing the passage of the liquid sodium.

Today, a technological maturity has been attained based on the following main choices: mixed oxide for the fuel, fertile blanket of depleted uranium oxide (in the breeder mode), stainless steel for the pin clads and the assembly wrapper tube.

However, other avenues are still being explored, such as carbide or nitride fuels, as is discussed below.

Fuel manufacture

FR fuel manufacture is similar to the manufacture of MOX fuel for LWRs, including high-temperature sintering of fuel pellets, using a mixture of powders of plutonium and uranium oxides. This fuel is in the form of solid pellets, annular pellets or vibro-fuel. Major differences are that the fuel stack is housed in a steel cladding tube, and that pin clusters are placed within a hexagonal steel wrapper tube.

The characteristics of the plutonium fuel entail the following special precautions:

- containment of the manufacture (in glove boxes) against α emissions;
- installation of biological shielding against γ-radiation and neutrons;
- protection against criticality risks; and
- protective measures to prevent diversion.

Table 12 lists the main manufacturing facilities, as well as the reactors that they have supplied. Manufacturing capacity was relatively low, and only the Cadarache plant in France had been designed to produce a large quantity of fuel intended for the manufacture of industrial scale Superphénix cores.

Table 12 **Fuel manufacturing facilities for fast reactors**

Manufacturing plants for FR fuel		Annual capacity (t) of heavy metal	FR supplied by this plant
USA	– Apollo, Babcock-Wilcox (ex NUMEC)	5 (a)	FFTF
Belgium	– Dessel, Belgonucléaire	5	SNR-300
United Kingdom	– Windscale, BNFL	5	PFR
Japan	– Tokai-mura, PNC	10	Joyo, Monju
Germany	– Hanau, SIEMENS	10 (b)	KNK-II, SNR-300
France	– Cadarache, COGEMA	20	Rapsodie, Phénix, Superphénix
Russia	– Chelyabinsk, Paket at Mayak	0.3	BN-350, BN-600
	– Dimitrovgrad, RIAR	1	BOR-60, BN-350, BN-600

(a)   now dismantled.
(b)   now permanently shutdown.

Spent fuel transport

Since the transport of new or spent FR fuel assemblies raises the same technical and regulatory challenges as the transport of MOX assemblies for thermal reactors, the transport facilities employed and the precautions observed are very similar.

In general, spent fuel assemblies are transported and dismantled at the reprocessing site: in this case, the cooling time before shipment is at least two to three years.

In any case, the assemblies are washed to remove the sodium and permit their storage in a pond in the reprocessing plant. Under certain circumstances, leaky or damaged assemblies are transported in special containers under inert gas or filled with sodium.

Spent fuel reprocessing

While the reprocessing of FR fuels makes use of the same process as the one used for thermal reactor fuel (the Purex process), a number of special factors must be taken into account: presence of sodium, use of stainless steel cladding and structure, high residual power, and high plutonium content.

While the experience gained is much smaller than that for reprocessing of thermal reactor fuels, France and the United Kingdom possess today solid experience in reprocessing FR assemblies. The FR fuel has been reprocessed either in specialised installations (e.g. the AEA plant at Dounreay in the United Kingdom, or the Marcoule site in France), or diluted with fuel from thermal power plants (La Hague, France). Part of the fuel from the Phénix and PFR prototypes was reloaded into the core after two successive reprocessings, thus demonstrating the complete fuel cycle.

Introduction

This section depicts, firstly, the present orientations and medium-term plans for the two countries, Japan and France, which pursue most actively the option of fast reactors. Further on, the perspectives on fast reactors in other countries are sketched.

Perspectives of fast reactors in Japan

The development of fast reactors in Japan has been so far based on Joyo and Monju.

The experimental reactor Joyo, which went critical for the first time in 1977 with a Mark-I core of an initial power of 50 MWt (later on it was increased to 75 MWt), was equipped in 1982 with a Mark-II core of 100 MWt. This core, in which the blankets are replaced by stainless steel reflectors, has been already operated for 29 duty cycles to test fuels and materials.

It is planned to raise soon the power again to 140 MWt with a Mark-III core; the refuelling times should be reduced in order to increase plant availability. The maximum fuel burn-up achieved so far in Joyo is 71 000 MWd/t (fuel pin average).

Monju, a prototype reactor of 714 MWt (280 MWe), went critical for the first time in April 1994. The first core of Monju was loaded with MOX fuel of a plutonium enrichment of 20 and 30 per cent in the inner and outer core zones, respectively. The core is surrounded by radial and axial blankets and is thus a plutonium breeder. At equilibrium, one fifth of the core fuel is to be discharged at each refuelling; the discharged fuel burn-up will reach 80 000 MWd/t (average). The successive core loads of Monju are planned to consist of the same MOX type fuel. Higher burn-up values will progressively be aimed at.

Japan has a programme of constructing a demonstration fast breeder reactor (DFBR) which would succeed Monju. Top-entry loop type primary cooling system, improved Monju type loop, and a 660 MWe (1 600 MWt) plant capacity were selected. The plant capacity is believed to be obtainable by extrapolating the Monju reactor physics performance to a 1 300 MWe size commercial plant. The construction will start in the early 2000s. Furthermore, the DFBR-2 plant, next to DFBR, is planned to be constructed at some appropriate time in order to make the LMFBR commercialised around 2030.

Perspectives of fast reactors in France and in Western Europe

The development of fast reactors in France is focused on Phénix and Superphénix. Research work has been important for nearly 30 years, not only in France but also in the neighbouring European countries Belgium, the Netherlands, Germany, Italy and the United Kingdom. The latter work, which was centred on DFR, KNK-II and especially PFR, has now been reduced to a very low level. The plan is to decide on building a European Fast Reactor (EFR) as a follower of Superphénix; this decision, now shifted after the year 2000, will depend on the lessons drawn from Superphénix. The EFR programme studies are now considering reduced plant sizes and the loading of plutonium burner cores.

Concerning the 250 MWe Phénix prototype, plans are to operate it up to the year 2005. The restart of Phénix to power has taken place in December 1994, after a four year interruption caused by inadvertent reactor shut-downs, but also utilised for a partial refurbishing of the secondary sodium circuits.

The major achievements on MOX fuel in the prototypes Phénix and PFR may be summarised as follows:

- pin cladding materials consisting of austenitic steels and nimonic alloys have been optimised and tested over high burn-up irradiations of more than 135 000 MWd/t in full sub-assemblies; in PFR, 180 000 MWd/t was even exceeded by experimental fuel; and

- ferritic wrapper tubes, virtually non-swelling under the impact of fast neutrons, have been developed to effectively reach these high burn-ups.

The 1 200 MWe Superphénix plant, owned by the European electricity utility NERSA, a joint venture gathering utilities from six countries, has been restarted in summer 1994 after a four year interruption initiated by a pollution of the primary sodium, which was followed by a public hearing set as a condition for the restart. Since its first criticality in 1985 and the start of full power production in 1986, the first core had, in 1990, experienced 174 equivalent full power (efp) days, to be compared to its target lifetime of 640 efp days.

In parallel with a public hearing, a commission called by the French Minister of Research H. Curien came to the conclusion that such a reactor was suitable to reduce, by MOX irradiation, amounts of accumulated plutonium.

The Curien report, therefore, recommended to use Superphénix as a plutonium burner instead of a breeder, which could be made possible by removing the fertile blankets (radial and axial) and increasing the enrichment. It further recommended to recycle in this reactor some amount of americium and neptunium, recovered from spent LWR fuel, to show that the reduction of these long-lived alpha-emitters can reduce the long-term toxicity of the nuclear waste.

In mid-1996, a second commission, headed by Professor Castaing, issued a report which essentially supported these recommendations.

In compliance with these new rules, the operation of Superphénix can be tentatively outlined as follows, up to the year 2010, which corresponds to a plant life-time of 25 years:

1994-1998:	Irradiation of the first core up to its core lifetime, with part of the radial blanket replaced by steel reflector;
1999-2002:	Irradiation of the second core up to the same burn-up target, with the whole radial blanket substituted by steel reflector;
from 2003 on:	Irradiation of the third core up to e.g. 960 efp days, with the fuel elements made of fissile section only; a fourth core load could follow.

While the two first core loads are available, the fabrication of the third one is still to be ordered. In terms of burn-up, 640 efp days correspond to 70 000 MWd/t (maximum) and 960 efp days to 100 000 MWd/t.

The third core (and the fourth core) could typically burn 200 kg plutonium a year, *i.e.* about the production of one LWR.

In these standard core loads, a number of specially optimised plutonium burner sub-assemblies, referred to as CAPRA elements, will also be inserted in a progressively increasing number; they belong to the CAPRA experimental programme which is described in Chapter 3.

During 1996, Superphénix went critical for 266 days, *i.e.* 95 per cent of the scheduled operation time. On receipt of the licences, its power has been progressively raised from 30 to 50-60 per cent of nominal power in February, and to 90 per cent of nominal power in the middle of October. The reactor was shutdown at the end of December, having reached 320 efp days, in order to perform a series of regulatory tests and core re-arrangements.

Perspectives of fast reactors in other countries

In the *Commonwealth of Independent States* (the former USSR), the operation of BN-600 in Russia and of BN-350 in Kazakhstan is a success, as was already outlined. Both reactors have basically been fuelled with enriched UO_2; plutonium test sub-assemblies containing so far more than 3 000 fuel pins have been successfully irradiated up to 100 000 MWd/t. This fuel, based on pellet technology, was produced at the Mayak plant at Chelyabinsk.

In parallel, the BOR-60 experimental fast reactor at RIAR (Research Institute of Atomic Reactors), Dimitrovgrad, has been loaded with vibropacked MOX fuel, manufactured on site up to an annual production capacity of one tonne of granulated fuel. So, BOR-60 was able to recycle its own plutonium.

The experience gained allowed to start the construction of:

- two fast reactors, BN-800 (of 800 MWe), on the Chelyabinsk and Beloyarskaya sites;

- a large MOX fuel manufacturing plant at Mayak RT-1 (Complex-300); and

- a new RT-2 facility to store and reprocess spent fuel from civil reactors and to fabricate MOX fuel.

However, financial difficulties have led to the suspension of the whole construction programme. Meanwhile, the BN-800 core was redesigned in order to increase plutonium consumption.

In the *United States*, the Fast Flux Test Facility (FFTF) of 400 MWt, which started in 1980 with a mixed oxide fuel core, had reached its burn-up target of 100 000 MWd/t for a full core load in 1987.

The Integral Fast Reactor (IFR) project, supported in the United States at the Argonne National Laboratory, has been based on the use of a metallic fuel for its core loading. Such a fuel, a ternary alloy U-Pu-Zr, was conceived as a successor of the alloy which has been recycled in the EBR-2 core after pyrochemical reprocessing of its spent fuel, with recovery of part of the fission products ("fissium"). Such alloys remain compatible with the steel cladding up to high burn-up values.

The IFR concept was integrated by General Electric into a full plant design called PRISM (Power Reactor Innovative Small Module), consisting of a series of reactor modules of 135 MWe each. The good inherent safety properties were stressed. The PRISM design was subsequently modified to raise the power level of the individual modules to about 300 MWe (PRISM Mod. B). All IFR designs were based on full actinide recycling using a pyrochemical processing plant collocated with the reactor complex.

At the beginning of 1994, the United States decided to stop the existing fast reactors (EBR-2, in operation since 1963, and FFTF), and to abandon the IFR project.

In *India*, mixed carbide (U-Pu)C fuel was loaded in the Fast Breeder Test Reactor (FBTR) near Bombay, and the first core of a very low power attained criticality in 1985. The second core should reach the power of 40 MWt; it requires about 200 kg of (Pu 0.55, U 0.45)C fuel pellets; the initial 15 kg of it have been produced so far (in 1993).

In *China*, the conceptual design of the First Fast Reactor (FFR), which begun in 1988, is basically completed. It should have a power output of 65 MWt/25 MWe, and could be in operation around the year 2000. Mixed oxide is foreseen as fuel. This design is supported by a R&D programme.

Irradiation performance, achievements and trends

Table 13 summarises the major achievements related to the irradiation performance of MOX fuel in fast reactors.

Table 13 **Irradiation performance of MOX fuel in fast reactors (major achievements)**

Country or group of countries	Standard (1) MOX fuel		Experimental fuel maximum burn-up reached MWd/t	Main reactors (3)	Type of fuel (3)
	No. of pins irradiated	Burn-up reached MWd/t			
Western Europe	265 000	135 000	200 000 (2)	Phénix, PFR, KNK-II	Solid and annular pellets
United States	64 000	130 000	180 000	FFTF	Solid pellets
Japan	50 000	100 000	120 000	Joyo	Solid pellets
CIS	13 000	135 000	240 000	BOR-60	Vibro-pacted
	1 800	100 000	–	BN-350	Solid and annular pellets
	1 500	100 000	–	BN-600	Solid and annular pellets

(1) The distinction between "standard" and "experimental" fuel is not obvious. "Standard" refers to the bulk of fuel pins comprised in full sub-assemblies and irradiated without special management measures.

(2) The figure of (approximately) 200 000 MWd/t of heavy material corresponds to pins loaded in PFR.

(3) In this summary, neither all reactors, nor all fuel types are quoted.

It is worth noting that the past fifty years of fast reactor fuel development have witnessed changes in popularity of the various fuel types from the initial use of oxide to emphasis on metal fuels, then to ceramics (mostly oxide) and finally back to both oxide and metal, as performance demands and priorities have changed.

Nevertheless, MOX fuel has dominated the scene in Europe and its excellent performance record is amply demonstrated in the pin statistics of Table 13.

It has been the cladding material rather than the fuel itself which has had the greater influence. The fuel material has presented few limiting factors, even when performance targets have been extended by a factor of three.

ATR

Introduction

The Advanced Thermal Reactor (ATR) is a heavy-water moderated, light-water cooled, pressure tube type reactor which is characterised by the capability to utilise plutonium, recovered uranium, etc. both flexibly and efficiently. It has been developed in Japan by using available domestic technology.

The prototype reactor Fugen (165 MWe) has been operating satisfactorily without any fuel failure since its commissioning in 1979. Experience has been gained in loading 602 MOX fuel assemblies (as of October 1995).

Plutonium utilisation characteristics of ATR

Since ATR is a heavy-water moderated, pressure tube type reactor, thermal neutrons required to sustain fission are mainly moderated in heavy water. Given that low energy thermal neutrons flow into fuel in pressure tubes arranged in square lattices, the effect of resonance absorption by plutonium and ^{236}U becomes small. Therefore, the ATR can use effectively MOX fuel with the same geometry and dimensions as UO_2 fuel, regardless of the plutonium isotopic composition vector. The ATR uses heavy water as a moderator which hardly absorbs thermal neutrons and it can achieve sufficient burn-ups with relatively low degree of plutonium and uranium enrichment. It can burn fissile material thoroughly.

Core reactivity is controlled by control rods which are placed in and out of the upper core and liquid poison (^{10}B) concentrations in the moderator. Control rods are mainly used in compensation for short time reactivity loss and for reactivity changes by external disturbance, and liquid absorber concentration in the moderator is used in compensation for reactivity loss by long-term fuel burning. Since control rods are inserted in heavy water, it is not necessary to change rod specifications and arrangements in using both UO_2 and MOX fuels.

MOX core characteristics

Since the core characteristics of the ATR are stable regardless of the fuels used, as mentioned above, the ATR can use several types of fuel: natural uranium MOX fuel, recovered uranium MOX fuel, enriched uranium fuel and so on. An example of the core configuration (22nd cycle) of Fugen is given in Figure 17. Figure 18 shows the history of MOX fuel utilisation in Fugen. Furthermore, ATR can use MOX in the whole core.

Mechanical and chemical characteristics of MOX fuel assemblies

No defective fuel has been found in Fugen fuel assemblies (FA), with design maximum burn-up of 20 000 MWd/t, for more than 15 years of normal operation and no abnormal oxidation, deformation, damage and so on has been found in post-irradiation tests. Fugen MOX FAs have a similar tendency in fission gas release rate and in metallographical tests of pellets as those of LWR fuel.

The maximum burn-up of MOX FAs has been increased to 40 000 MWd/t and the number of fuel elements is also increased from 28 cluster to 36. A further 54 cluster-fuel which is aimed at high burn-up (55 000 to 60 000 MWd/t) is under development.

2.7   Reprocessing of MOX fuel

Specificities of MOX fuel reprocessing

In order to achieve the principal goal of the reprocessing and recycling strategy, *i.e.* the maximum recovery of valuable materials, it is important to be able to reprocess MOX fuels. This is a necessary step towards multi-recycling.

Spent MOX fuel is similar to spent UO_2 fuel, except from a different plutonium content and a different nature of matrix (produced by blended PuO_2 and UO_2 powders).

The first feature can be dealt with in today's reprocessing plants. The second feature results to some differences in the solubility of MOX fuels, as compared to UO_2 fuels. Research and Development (R&D) experiments and industrial scale reprocessing showed that the feasibility of MOX dissolution is very close to that of UO_2 fuels.

Finally, uranium/plutonium partition and waste management are very close to those of standard fuels, and could be performed easily in current reprocessing plants, with the same level of performance as for UO_2 fuels.

R&D programmes on MOX fuel reprocessing

Laboratory investigations of MOX fuel dissolution

The French experience with mixed oxide fuel reprocessing actually began in 1967 with fuels from fast breeder reactors (first Rapsodie and then Phénix). The German experience included, in the late 1970s, reprocessing MOX fuel from a PWR (KWO) and an HWR (MZFR) at an experimental reprocessing facility in Karlsruhe, refabricating therefrom MOX fuel at the Alkem plant in Hanau and loading it in the KWO PWR. The Japanese experience began in late 1982 in the Chemical Processing

Figure 17.   **Core configuration of Fugen (22nd cycle)**

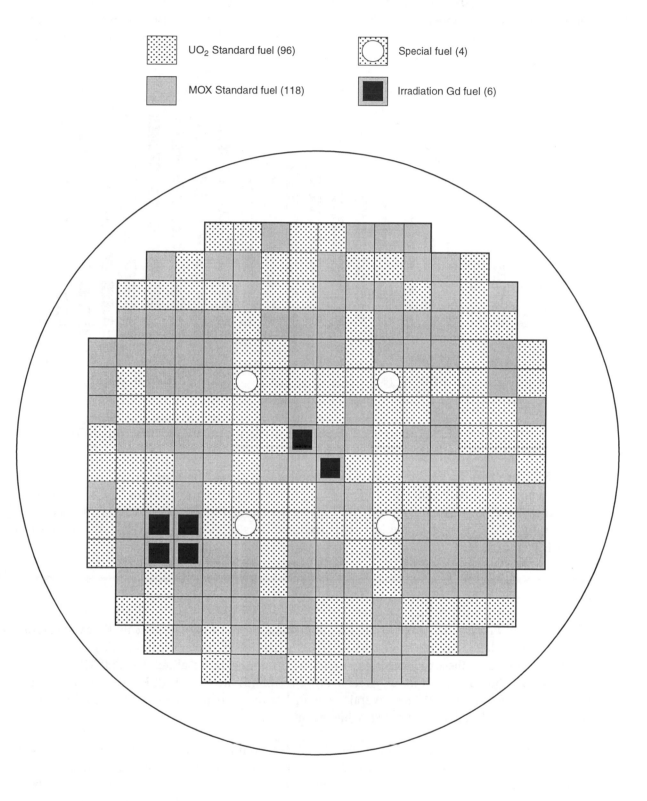

Figure 18. MOX fuel utilisation in Fugen from March 1978

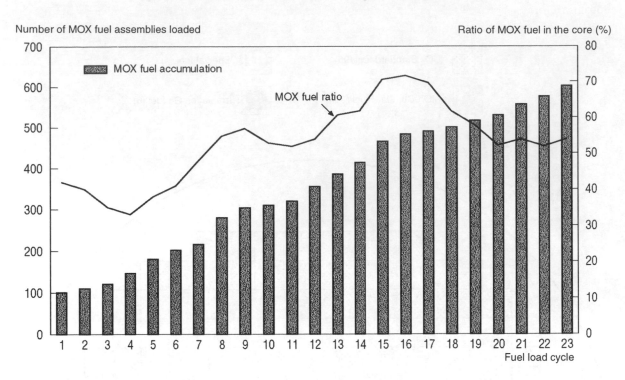

Number of MOX fuel assemblies loaded

Ratio of MOX fuel in the core (%)

Facility (CPF) devoted to FBR fuels reprocessing. It has been shown that dissolution of the plutonium of those mixed oxide fuels in nitric medium depended mainly on:

- The fabrication of the ceramics: dissolution is very sensitive to the composition of the powders and to the heterogeneities of the mixed oxide (UO_2 and PuO_2).

- Their irradiation: on the one hand, nuclear reactions consume initial plutonium and generate new one, leading to a more even plutonium distribution like in UO_2 fuels. On the other hand, the irradiation and the high temperature favour the diffusion of plutonium through the UO_2 matrix; the last phenomenon differs depending on whether the reactor is a FBR or a LWR.

Japanese experiments

The Chemical Processing Facility (CPF), comprising a series of miniature scale Purex apparatus, entered operation in late 1982 for laboratory scale hot experiments on reprocessing FBR MOX fuels. MOX fuels dedicated to the experiments had reached a burn-up up to about 100 000 MWd/t in the experimental fast breeder reactors Joyo (Japan), Phénix (France) and PFR (United Kingdom). Through experience gained with the several kinds of hot experiments in the CPF, recycling of the plutonium product to the Joyo reactor was achieved in 1984.

Results obtained up to now indicate that the fuel dissolution was very satisfactory. The key parameters affecting the rate have been defined. The amount of the insoluble residues and its composition were evaluated by changing the dissolution conditions. This information has been successfully reflected in the design of the continuous dissolver and its operating conditions in the RETF (Reprocessing Equipment Testing Facility). The application of Purex process data to FBR fuel reprocessing design has also been achieved. For instance, the chemical behaviour of uranium, plutonium, minor actinides and fission products was clarified. The separation of these elements with a sufficient decontamination factor (DF) was obtained in a minimum number of extraction cycles.

As mentioned above, R&D in CPF has covered a wide range of technical topics, and more advanced studies to develop new reprocessing technologies will be continued in the future.

French experiments

In FBR reactors, the fuel temperature is higher than the oxide pellet sintering temperature (usually 1 700°C). The fabrication heterogeneities are "erased" and disappear during irradiation.

On the contrary, with MOX fuels for LWR reactors, in which the fuel temperatures never exceed 1 300 to 1 500°C, it is necessary to find the degree of heterogeneity that could be tolerated during their fabrication in order to be able to be "erased" and hence, subsequently, reprocessed without any particular difficulty. In this context, studies have been undertaken since 1987 in hot cells at Fontenay-aux-Roses.

Dissolution and solubility

Table 14 shows the results of solubility measurments obtained with French and German MOX fuels, before and after irradiation. After irradiation, each dissolution test was performed on the whole rods (1.5 to 2 kg U and PuO_2), sheared into 35 mm pieces, in industrial dissolution conditions. Before irradiation, the measurements were taken on the pellets.

Table 14 **Solubility measurements on PWR-MOX fuels manufactured by mechanical blending**

Used fuel			Maximum final insolubility*	Complementary disolution 10 N-12 h at boiling **
Source of fuel	Total Pu content	Burn-up (GWd/t)		
France	5%	27	0.028%	n.d.
Germany (I)	5%	30	0.024%	65%
Germany (II)	4.7%	34	0.028%	80%

*   % of plutonium undissolved relative to the plutonium present in the fuel.
**   % of initial insolubles dissolved.

After irradiation, it was experimentally confirmed that the dissolution of MOX fuels was much more complete than before irradiation.

For a burn-up limited to 30 GWd/t (at equilibrium it is planned to irradiate these fuels to 45 GWd/t), the MOX spent fuel plutonium solubility proved to be very high, at around 99.97 per cent in all the measurements; whereas, before irradiation, the average solubility was 99.5 to 99.6 per cent.

As shown in Table 14, processing the dissolution residues (dissolution fines) by complementary boiling attack in 10 N nitric acid over periods of 6 to 12 h, leads to a recovery of up to 80 per cent of the remaining plutonium in these fines.

Dissolution wastes

The dissolution fines from MOX fuels are closely similar to those of UO_2 fuels. For the same burn-up (33 GWd/t), the total mass of fines ranges from 4.0 to 4.5 kg per tonne of uranium and plutonium.

Apart from the fact that they may contain PuO_2-rich particles, their composition is roughly similar. The actual mass composition depends on the irradiation history, which is specific to each treated fuel, and to a large extent on the dissolution conditions which dictate the solubility of Mo, Zr and Sn. Their maximum plutonium content can reach about 0.3 per cent.

Industrial scale treatment

British industrial experience with fast reactor fuel reprocessing

The fast reactor reprocessing plant at Dounreay was based upon the earlier Dounreay fast reactor reprocessing plant. A new fuel disassembly cave, waste facilities and a plutonium production section were commissioned in 1980 and the plant commenced operation in the autumn. Since that date, the plant has reprocessed over 23 t HM discharged from the Prototype Fast Reactor (PFR). (The PFR was definitively shut-down in March 1994, after 20 years of satisfactory operation.)

The plant uses a modified Purex process originally developed at Hanford, USA: a three cycle flow-sheet with final purification of the plutonium nitrate product. The design of the plant is based on safe-by-geometry mixer settlers.

The dissolver basket contains, typically, 128 pins with between 4.5 and 5.5 kg of plutonium and between 23 and 25 kg of uranium. The dissolver cycle time is less than 24 hours and it is the rate limiting operation which is determining the plant throughput to a maximum of around 5 kg of plutonium per day for PFR core fuel. The dissolution of irradiated fuel has proved to be efficient and losses of plutonium to insolubles were lower than predicted. Since 1980, the Dounreay plant extracted over 3.5 t of plutonium. This plutonium nitrate solution has been sent to Sellafield for conversion to oxide.

The reprocessing of fast reactor fuel makes use of the same process as that for thermal reactor fuel, but two special factors have to be considered:

- Removal of sodium

  The metal used as coolant in the PFR is sodium. It presents certain drawbacks: at operating temperatures, it ignites spontaneously, if in contact with air and its contact reaction with water is violent. In the fast reactor reprocessing plant, the sub-assemblies are steam cleaned to remove any residual sodium.

- High plutonium content

In the dissolver, typically, more than 99 per cent of the plutonium and 100 per cent of the uranium go into solution. The residual particulate plutonium is removed by centrifuge. The centrifuge bowls are changed after each dissolution and transferred to long-term storage as Intermediate Level Waste (ILW).

The fast reactor fuel reprocessing plant is subject to IAEA and EURATOM safeguards. Inspections are carried out by EURATOM.

French industrial MOX fuel reprocessing

Nearly 28 t of FBR MOX fuel have been reprocessed in different French installations (mainly Marcoule and La Hague), without any particular problems, and with dissolution yields greater than 99.8 to 99.9 per cent. Currently, the Marcoule plant is reprocessing Rapsodie cores. This section focuses on the industrial reprocessing of thermal MOX fuel in France.

MOX fuel reprocessing at APM

The Marcoule pilot facility (APM, Atelier Pilote de Marcoule, in the south of France) conducted a semi-industrial campaign in early 1992 by reprocessing 2.1 t of MOX fuels.

Table 15 shows the characteristics of the reprocessed MOX fuel before and after irradiation.

Table 15 **Characteristics of MOX fuel reprocessed at APM**

Before irradiation		After irradiation	
% fissile Pu:	2.0 to 3.2%	% fissile Pu:	about 2%
Fabrication :	co-milling	burn-up:	about 34 GWd/t
Solubility :	about 99.6%	Cooling time:	3.5 years

The dissolution kinetics observed at APM confirmed the laboratory results.

Regarding the material balance of uranium, plutonium and main fission products, consistency was shown between the results of the computer codes and the experimental results. The discrepancies observed in the dissolution liquor for a few fission products (^{106}Ru, ^{125}Sb) can be explained by their poor solubility: these radionuclides are well known components of the dissolution fines.

Concerning the obtained performance, the residual contamination figure for the hulls was close to the laboratory test values produced during the laboratory investigations.

The dissolution fines recovered after clarification contained a small plutonium amount (about 0.2 per cent of the plutonium present in the fuel). An additional attack test has been performed on these fines which led to recover less plutonium than expected according to laboratory tests.

The dissolution liquor was treated without dilution on the APM extraction line. Plutonium losses in the extraction raffinates and the unloaded solvent were small. The uranium and plutonium nitrates were concentrated separately. The characteristics of the end products showed that the specifications were met with only two purification cycles.

MOX fuel reprocessing at the UP2-400 plant at La Hague

Following the conclusive experiment with MOX fuel reprocessing at the Marcoule pilot facility, COGEMA conducted an industrial reprocessing run in November 1992 with 4.7 t of MOX fuel in the UP2-400 plant at Cap La Hague.

Table 16 shows the characteristics of the reprocessed MOX fuel before and after irradiation.

Table 16 **Characteristics of MOX fuel reprocessed at UP2-400**

Before irradiation		After irradiation	
% fissile Pu :	up to 3%	% fissile Pu:	about 2%
		burn-up:	33 to 41 GWd/t
Solubility :	99.6 to 99.8%	Cooling time:	about 5 years

This fuel was representative of the MOX fuel currently supplied by fuel manufacturing plants.

The reprocessing conditions were as follows: the MOX fuel was dissolved in 6 N nitric acid (final acidity about 4 N) and the average residence time of the hulls in the batch dissolver was four hours.

The flow-sheets adopted for the purification cycles were similar to those usually applied for UO_2 fuel. These flow-sheets were designed using the process models developed by the CEA. The dissolution liquor was diluted in reprocessed uranium before the first extraction cycle in order to adjust the plutonium/uranium ratio to 2 per cent. (In the future, MOX fuel will be diluted – when reprocessed – with UO_2 fuel by a ratio of roughly 1/4.) The average treatment rate was around one tonne per day in the different facilities.

Residual contamination of the hulls, after water rinsing, and plutonium losses in the purification cycles (extraction raffinate and unloaded solvent) were similar to those usually observed in the plant. Hence, the higher plutonium flows due to the MOX fuel did not alter the performance.

Special analyses were performed on dissolution insolubles separated at the clarification step. The mass of dissolution fines was estimated at about four kg per tonne of uranium and plutonium. Their plutonium content was low (about 0.1 per cent by weight), corresponding to about 0.01 per cent of the plutonium present in the irradiated fuel.

All the specifications on the uranium and plutonium end products were met.

Summary

Experiments conducted in French and Japanese laboratories with MOX fuel showed that irradiation in LWR reactors led to erasing most of the homogeneity flaws of the plutonium in the tested MOX fuels. The final solubility of plutonium was, consequently, very good (more than 99.97 per cent) and had no effects on the process conditions.

The feasibility of reprocessing MOX fuel as manufactured today was demonstrated on semi-industrial scale in APM, and on industrial scale in the UP2 plant at La Hague, in conditions similar to those employed for reprocessing UO_2 fuel. MOX fuel reprocessing is not necessary in the next few years, as utilities recycling plutonium have enough material to fabricate the necessary MOX fuel.

The technological options applied in the current industrial reprocessing plants for reprocessing MOX fuel were then accordingly confirmed. The way to multi-recycling is, thus, open today and plants have already separated second generation plutonium for future use in MOX fuel.

2.8   Plutonium purification

Introduction

Goals of plutonium purification and advanced recovery

An efficient plutonium management scheme must deal with plutonium ageing and americium accumulation, if needed. After the year 2000, it is expected that the available capacity for MOX fuel fabrication will allow transferring efficiently the plutonium output from reprocessing plants directly into the MOX fuel fabrication facilities without long storage periods. Nevertheless, the ability to purify plutonium is an important factor of flexibility since it ensures, even in the case of temporary fluctuations, that "old" plutonium oxide powder can be cleaned from americium.

As much as possible, the overall waste activity should be limited. As a result, it is necessary to minimise the amount of plutonium in waste sent to disposal together with the various kinds of wastes generated during the plutonium processing operations. The implementation of efficient waste treatment processes able to recover plutonium from a wide range of materials guarantees that the amount of plutonium disposed of would be extremely low, even in the case of major evolutions in the recycling technology.

In countries practising recycling, the above principles have been included early into the design of the back-end of the commercial plutonium cycle, resulting in the management concept outlined in Figures 19 and 20 (which represent the French case). As it can be seen from these figures, in the main plutonium flow leading from reprocessing to the fabrication of MOX fuel, two specific loops have been designed for:

- purification and recycling of non-directly re-usable material; and

- recovery and recycling of plutonium from waste generated during the operation and maintenance of plutonium processing facilities, in association with incineration, in order to reduce both the volume and the activity of the waste.

This concept takes advantage of existing reprocessing facilities for the recycling of the plutonium thus recovered. It has also been designed to take care of the plutonium-bearing waste and scraps that had been generated and previously stored.

Figure 19. **Management of "old" plutonium and "scraps"**

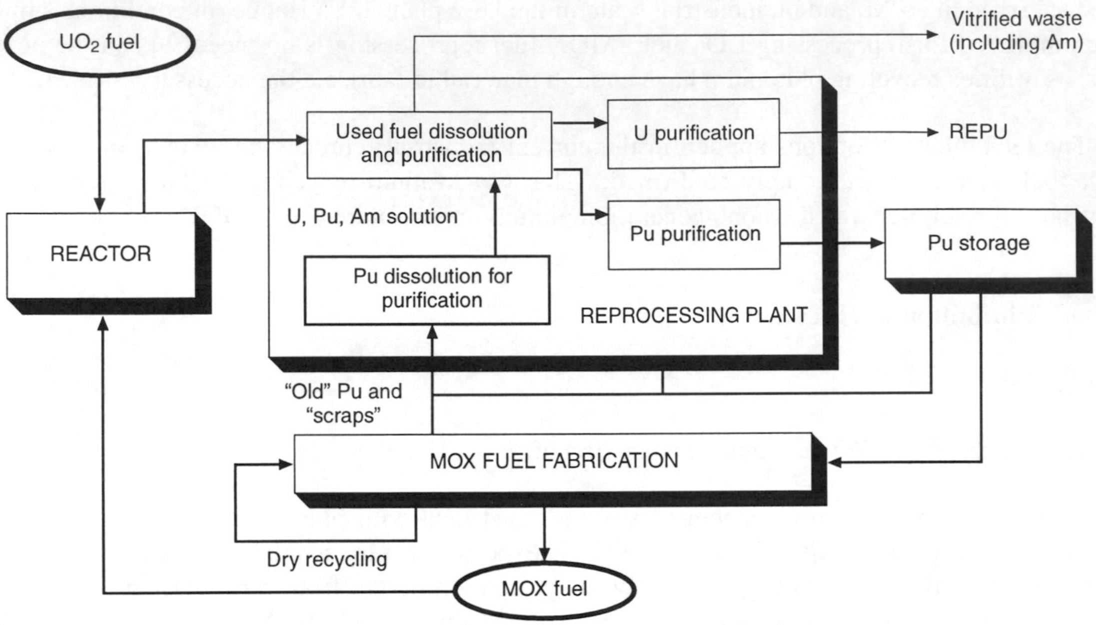

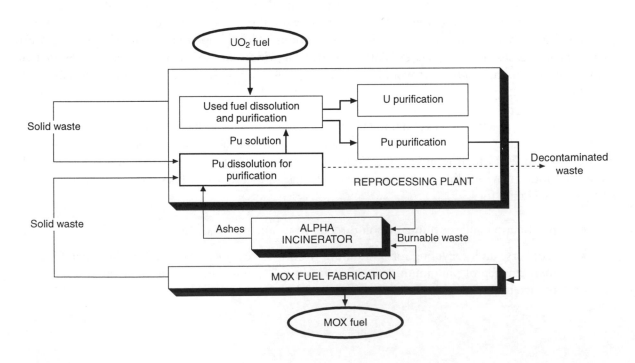

Figure 20. **Plutonium recovery from waste**

Review of the techniques

In the present state of the plutonium cycle, the materials to be dealt with in the side loops, described above, may arise from all operations, from reprocessing to fresh MOX fuel delivery and may be classified into the following:

- "Scraps" from MOX plants and "old" PuO_2 which are, in fact, raw materials (PuO_2 or mixed UO_2 and PuO_2 in the form of powder or pellets) not suitable for direct use in the MOX fabrication plant.

- "Contaminated waste", where plutonium is only present as a contaminant, more or less abundant, more or less fixed to a substrate. Such waste may be directly issued from the plutonium-processing facilities (metal waste, for instance), or be the result of the incineration of plutonium-rich burnable waste (some plastics, cotton, wool, organic materials, etc.).

The following sections describe two very versatile electrolytic recovery techniques dealing with these two types of material. The corresponding facilities are able to process the "secondary" flows of plutonium induced in any plant managing plutonium.

These techniques are in industrial operation today in France. In addition, they are extensively studied in several other countries all over the world: Russia, Japan (JAERI, PNC), the United Kingdom and the United States (Lawrence Livermore, Los Alamos). Moreover, the principle of oxidative dissolution with electrogenerated mediators can be extended to other applications, among which the destruction of contaminated organic materials. This last application is, in particular, studied in the United Kingdom (Dounreay) and in the United States (Savannah River, Lawrence Livermore).

In France, the last large-size facilities needed to perform the plutonium recovery operations have been commissioned in 1994 and are operating under the international safeguards regime of IAEA and EURATOM. The main result is that, by now, it is possible to reduce the proportion of plutonium sent to disposal with the waste to an order of magnitude of 0.1 per cent of the total plutonium content of the spent fuel. The examples given hereafter are based on the French experience.

Plutonium purification and management of "scraps"

The principle

For "old" plutonium, as well as for "scraps", recovery consists in dissolving the incoming material (PuO_2 powder, UO_2-PuO_2 powder or UO_2-PuO_2 pellets) in nitric acid and purifying the resulting plutonium and americium or uranium and plutonium solutions.

The plutonium produced in reactor fuel is mainly composed of five isotopes: ^{238}Pu, ^{239}Pu, ^{240}Pu, ^{241}Pu, ^{242}Pu. All these isotopes undergo spontaneous decay. ^{241}Pu emits a beta-particle to produce ^{241}Am (which has a short half-life of 14.4 years). ^{241}Am is a gamma emitter.

When plutonium is made available to a MOX fuel manufacturer, ^{241}Am increases the personnel exposure in the MOX fabrication plant. A maximum level of ^{241}Am for plutonium entering the plant or at the beginning of the fuel fabrication processes lies between 10 000 µg per g of plutonium for the old plants and 30 000 µg per g of plutonium for the new ones. Therefore, "old" plutonium and "old"

101

MOX "scraps" have to be purified, *i.e.* americium must be removed, unless the old plutonium can be diluted by fresher plutonium.

In addition, [241]Am is a neutron absorber. With time, it decreases the fissile worth of the plutonium. Removal of the americium gives a higher energetic value to the purified plutonium.

The purification of nitric plutonium and/or uranium-bearing solutions is routinely performed as part of the reprocessing operations. Most reprocessing plants use the Purex solvent extraction process, with many variants: for instance, in the RT2 facility being constructed near Krasnoyarsk (Russia), the first cycle will involve simultaneous recovery of uranium, plutonium, neptunium, technetium and zirconium, with subsequent separate re-extractions of zirconium, plutonium, neptunium, technetium and uranium. Uranium and plutonium will then be purified in specific cycles (47). At La Hague, uranium and plutonium are separated from the bulk of fission products and stripped from contaminants (zirconium, technetium) prior to partition and further purification in separate cycles (48). The fission products and contaminants are vitrified as high-level waste. Consequently, the French approach for plutonium recovery from "old" plutonium and "scraps" consists of making compatible and injecting the nitric plutonium and americium or uranium and plutonium liquors resulting from dissolution at selected locations of a reprocessing plant. With this method, americium and other contaminants brought in by the dissolved plutonium materials are efficiently removed and directed to vitrification together with the high-level waste, while the recovered uranium and plutonium re-enter their respective cycles through the normal routes (see Figure 19).

The major difficulty to overcome for the recovery of PuO_2 or UO_2-PuO_2 materials was the dissolution step, since the conventional technique used at the front-end of the reprocessing plant for irradiated fuel (mere contact with hot concentrated nitric acid) is not efficient for the dissolution of non-irradiated, solid PuO_2 particles. It was, therefore, necessary to develop an alternative method compatible with the industrial environment for the Purex process.

Development of the process

The principle of oxidising dissolution for PuO_2 was first proposed in 1975 by HORNER et al. (49) from Oak Ridge National Laboratory and HARMON (50), using Ce(IV) as a mediator. In 1981, BRAY and RYAN (51) from Pacific Northwest Laboratories proposed a dissolution method using electrogenerated mediators such as Ce(IV), Co(III) or Ag(II). Among the various processes studied world-wide, the PuO_2 dissolution method which has been selected is based on the oxidation of insoluble solid Pu(IV) (PuO_2) to soluble Pu(VI) (PuO_2^{2+}) species in a nitric medium. The oxidation is performed via an electrochemical mediator with a strong oxidising power: Ag(II) (Ag^{2+}). This mediator is continuously regenerated by application of a fixed current to the platinum anode of an electrolyzer. The amount of americium that may be present in the material is also dissolved by this process.

From the beginning of the eighties, KOEHLY, BOURGES, MADIC and LECOMTE (52) from the Commissariat à l'Energie Atomique, performed a systematic parametric study of the oxidising method using silver and showed that the rate limiting factor was the rate of generation of Ag(II). From these studies, a pilot was built in 1985 at Fontenay-aux-Roses which demonstrated the feasibility of quantitative dissolution of PuO_2 and allowed optimising the operating conditions. The pilot included the main features of the future industrial units: basic conception of the electrolyzer, control of the reaction using on-line spectrophotometry to measure the concentrations of Pu(VI), Am(IV) and Ag(II). This pilot is able to dissolve batches of about 850 g of plutonium and americium oxides in three hours with an electrolysis current of 80 to 100 A.

Figure 21. **Industrial dissolver for "old" PuO$_2$ and "scraps"**

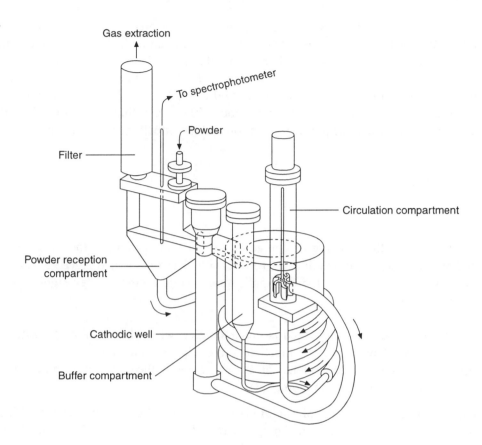

Industrial application

An example of small-size unit: the T4 facility in UP3

In 1989, a unit very similar to the pilot and able to dissolve batches of one kg of PuO$_2$ powder in four hours, has been commissioned in the T4 facility of the UP3 plant at La Hague to recycle the remnants of PuO$_2$ samples and dust recovered from the T4 PuO$_2$ processing facility itself. At the end of the dissolution operation, the dissolution liquor is made compatible with the Purex process and is recycled back to the last plutonium purification cycle of UP3.

An example of large-size unit: the URP facility

More recently (1994), the centralised large capacity industrial unit able to perform the recycling of "old" plutonium and of the "scraps" from MOX fuel fabrication has been commissioned in the R1 facility of the new UP2-800 reprocessing plant at La Hague. This unit - URP - is able to process PuO$_2$ powder, as well as mixed UO$_2$-PuO$_2$ powders or sintered pellets from various origins. Its annual capacity ranges from four tonnes of (PuO$_2$ and AmO$_2$) powder to up to nine tonnes of mixed oxides, according to the nature of the feed and the rate of operation of the reprocessing plant. It uses a new generation of dissolver with high current intensity (400 A), optimised current yield, and increased subdivision of the various functions into elementary steps performed in geometrically safe zirconium compartments as shown on Figure 21. The dissolution liquor is diluted and recycled, either at the head of the plant (for uranium and plutonium) or at the entrance of the plutonium purification line (for plutonium and americium).

The electrolysis vessel is fit with a platinum anode (where the regeneration of Ag(II) ions occurs) dipping in the PuO_2 nitric suspension and a porous cathodic well in which the catholyte circulates. The cathode is made of tantalum. The slab-shaped powder reception compartment is connected to the mechanical powder supply systems. In the circulation compartment, an impeller circulates the solution and keeps the powder dispersed in the liquor. The buffer compartment allows adjusting the final volume for a concentration of about 250 g/l. Cooling (or heating) is performed in a coiled jacketed circuit between the powder reception and the impeller compartments.

Plutonium recovery from solid waste

The principle

A method based on the leaching of contaminated waste according to the same principle as the one described above – oxidative dissolution of PuO_2 with electrogenerated Ag(II) as a mediator – can be used. However, the conditions for the application of this method differ significantly from the dissolution of pure oxides:

- The mass of plutonium involved is much smaller, while the volume of waste to process is rather large. This results in much larger equipment and the necessity to optimise the circulation of the solution to leach all the waste, while the concentration of plutonium in the solution may be quite dilute (some g/l).

- The substrate on which the plutonium is (more or less) fixed may interact with the solution and, thus, consume some of the energy provided and introduce undesirable chemicals in the solution. As a result, the current yield is much lower and it is necessary to filter and purify the leaching liquor prior to recycling in the main reprocessing plant.

- Pre-treatment may be useful to ease the removal of plutonium from some substrates: for instance, for some burnable waste, it is much easier to dissolve plutonium from incineration ashes than from the waste itself. As a result, incineration (which also reduces the volume) is applied as much as possible prior to decontamination.

Development of the process

The experience of CEA on active and inactive pilots has demonstrated the efficiency of the method for a large variety of waste: up to now, the PROLIXE shielded, remote-handled facility commissioned in 1988 at Fontenay-aux-Roses has decontaminated over 300 kg of metallic waste down to low level (< 0.1 Ci/t α after 300 years). For non-irradiating α-waste, the ELISE glove box facility, commissioned in 1987, performs the same function. The treatment of several kg of PuO_2-rich ashes in the CASTOR active facility allowed to define operating guidelines for a plutonium recovery yield of 97 per cent or more. The TRACEM inactive pilot for ashes is used for industrialisation studies at Marcoule.

Industrial application

An example of small-size unit: the T4 facility in UP3

The T4 facility is the part of UP3 that processes plutonium. This facility generates plutonium-bearing waste which, for a large part, belong to the category of low-level waste. A small size unit based on the silver dissolution process, which is manually operated in glove boxes, has been commissioned for the specific use of T4. The 40 l electrolyzer is able to process annually around 100 batches of stainless steel filters, failed metal equipment and some plastics.

An example of large-size unit: the UCD facility

This unit, located in the R2 facility of UP2-800 at La Hague has been designed to process, in two parallel leaching lines, solid plutonium-contaminated waste from all La Hague facilities (except T4), from the MOX fuel fabrication plants and from storage facilities where waste had been stored since the start of the La Hague reprocessing activities. A third line is dedicated to the processing of incineration ashes and dust from the COGEMA's α-waste incinerator located near the MELOX facility.

In the two solid waste leaching lines the solid waste (metal cans, some plastics) are received in containers, unloaded, sorted in batches according to the type of material, submitted, when necessary, to mechanical pre-treatment (cutting, grinding, etc.), leached in a specific, conventional geometry 200 l titanium tank by Ag(II) ions generated in a 400 A electrolyzer, rinsed and dried prior to unloading, control and conditioning. Operating parameters are specified for each type of waste. About 800 batches can be treated per year.

In the third line, the ashes are received in containers, unloaded, crushed and dechlorinated by electrolysis prior to injecting the Ag^+ catalyst. The fissile material and a part of the ashes are then dissolved by the silver method in a three compartment geometrically safe, 400 A, titanium electrolyzer. The resulting suspension is filtered. The plutonium-containing solution is recovered and the decontaminated pulp is routed to vitrification. The size of a batch is about 6.6 kg. The annual capacity is about two tonnes of ashes.

The solutions recovered from the three lines are purified from undesirable chemicals in a specific cycle prior to being dispatched to the head of the reprocessing plant. Silver is recovered by electrolysis from the raffinates of this purification cycle and recycled back into the electrolyzers.

Summary

The above examples show that the purification of plutonium and its recovery from ashes and solid waste are now practised at the industrial scale. The implementation of such operations in the plutonium cycle provides two major advantages to the resulting system:

- improved flexibility of the overall plutonium management since it would allow for purification of "old" plutonium that has accumulated significant amounts of americium; and

- very small levels of plutonium in the ultimate waste.

Chapter 3

FUTURE DEVELOPMENTS

3.1 Introduction

The future utilisation of plutonium will critically depend, among other factors, on the evolution of the nuclear policies. This chapter addresses plutonium use in fast and other reactor systems, as well as plutonium conditioning to a form compatible for final disposal. Such technologies, currently under research and development, would need to be fully demonstrated and accepted.

In evaluating strategies for the future utilisation of plutonium, several issues need to be assessed, relating to safety implications and to political, economic, technical and ethical conditions. In particular, utilisation of plutonium should be carried out in accordance with non-proliferation and environmental requirements.

A main concern for the next generation of reactors will be managing nuclear materials and the associated waste in a satisfactory way. The reactors should ensure an optimum plutonium consumption and produce a minimum amount of actinides. Plutonium recycling in light-water reactors is a necessary but incomplete step. Fast reactors will allow the fuel cycles to be closed effectively and will better satisfy waste minimisation requirements, both as consumers of plutonium in a transient phase and as breeders in the long term.

Currently, there exist surplus plutonium stocks which could be reduced by continued utilisation of MOX in thermal reactors. This option can be implemented by using the current generation of fuel fabrication plants. Such an approach may include modifications to reactors in order to increase the rate of plutonium consumption, particularly if large amounts of non-civil plutonium are transferred to the civil sector, either by allowing higher MOX contents in light-water or CANDU reactors, or by substituting the fertile zone of a fast reactor by a reflector.

In the medium term, *i.e.* after two decades, a mixed nuclear power production park composed of LWRs and Fast Reactors (FRs) could stabilize the plutonium and minor actinide (MA) inventory, and minimise waste, especially TRU waste, production. The degraded plutonium spectrum resulting from LWR-MOX irradiation limits the recycle capability in thermal reactors and, hence, FRs would be needed to consume the residual fraction of spent LWR-MOX fuel. Some recycle of MAs would, in principle, be possible. However, MA recycling would have important consequences for fuel cycle facilities which separate MAs from fission products and fabricate them as MOX fuel.

Concerning the longer term, very efficient, reliable and safe fast breeder systems could be envisaged to extract all the fissile energy from the stockpile of depleted uranium which would have been accumulated over the years throughout the world. These systems may incorporate modular and integral type reactors with metal, nitride or carbide fuel. They could operate with equilibrium fuel cycles using the uranium stockpile without increasing the overall TRU inventory, minimising, in this way, the waste and TRU generation per unit of energy produced.

107

Should it be decided by national policies not to use plutonium as a fissile resource, stocks may in future be conditioned to a form compatible with regulatory requirements for final disposal. Deep borehole disposal may also be considered.

The nuclear fission era might be terminated with the introduction of accelerator-driven transmutation systems, if the public and the authorities of future societies consider the disposal of large amounts of spent TRU, and in particular plutonium, as not being appropriate.

3.2   Plutonium burning in current reactor concepts

Introduction

There are various R&D activities going on in several countries regarding the utilisation of reactors, currently used for power generation, for managing present and future plutonium inventories.

As designed, a fast breeder reactor produces more plutonium than it burns, but, at the conceptual stage, it is possible to transform it into a burner if reducing the plutonium inventories is needed, whether the plutonium originates from power reactors or from non-civil uses. Similar concepts can also apply to CANDU reactors and LWRs.

In France, a study named CAPRA project has been launched to demonstrate the feasibility of a plutonium-burner fast reactor core compatible with the European Fast Reactor (EFR) technology. Some calculations have been made to estimate a rough figure for Monju, which could annually burn a total amount of about 80 kg of plutonium. For reactors like BN-600 and Superphénix with much higher thermal power than Monju, correspondingly higher net burning yields could be expected. Similar studies have also been carried out in Japan.

In Japan, two conceptual studies have been carried out by JAERI on the modification of high conversion PWRs in reducing surplus plutonium and on a new concept for a LWR fuel matrix which can become chemically stable during irradiation, hence capable of being disposed of without further processing. The modified PWR, called an active storage PWR, would have a reactor core with tight triangular pitched lattices and a lower moderator-to-pellet volume ratio. The new LWR fuel concept has a chemical composition that is basically the same as that of a mineral with the fluorite type structure, so that actinide and lanthanide elements, typical fission by-products, can be dissolved in the structure, and can be fabricated in a conventional facility by conventional methods. A similar programme of burning plutonium in LWRs is being studied also in Europe.

CANDU

Although CANDU reactors have not been fuelled with MOX fuel to-date, they have been operating with natural uranium in which fissile plutonium is the largest contributor to energy production. This feature is a result of the high conversion ratio of the natural uranium/heavy-water lattice. Over 200 reactor-years of operation demonstrate the ability of existing CANDU plants to burn plutonium-bearing fuels.

The need to disposition large quantities of separated plutonium has provided an incentive to increase the plutonium content of the fuel to be used in the CANDU-MOX fuel cycle. The MOX fuel cycle utilises a standard CANDU fuel bundle that has 37 pins or elements that are arranged in three rings around a central pin. The outer rings contain MOX fuel pellets with weapons or reactor-grade plutonium dioxide that is mixed with depleted uranium dioxide. The plutonium concentration is graded, with the outermost ring having a composition of about 1.4 and the second ring about 2.3 wt. per cent total plutonium. The innermost ring and central element contain a mixture of depleted uranium dioxide and 5 wt. per cent dysprosium oxide. The dysprosium oxide functions as a burnable poison which matches the reactivity decay of a MOX lattice in CANDU.

Relevant features of the MOX-fuelled CANDU plant are as follows:

1.   The annual throughput of plutonium is 1.7 t per GWe of installed CANDU capacity. The fissile plutonium content of the bundle is reduced by 60 per cent on discharge.

2.   The fuel operates within the existing envelope of experience gained with natural uranium fuel. Fuel behaviour, including element stresses and strains and gaseous fission product formation are such that the reference fuel design meets all existing requirements. Evaluations of the fuel's thermal-hydraulic behaviour, including determination of critical heat flux show that the existing margins to fuel dryout are maintained.

3.   The rate of fuelling of the reactor core is slightly lower than at present, so that no modifications are required to the on-power fuelling machines.

4.   The stability and control requirements of the MOX-fuelled CANDU reactor are unaltered from the current operating reactors that use natural uranium as, due to the high conversion ratio, a large fraction of the energy produced with natural uranium is from plutonium. As expected, the feedback reactivity co-efficients are essentially unchanged (Table 17). The changes in neutron kinetics parameters are small so that no modifications are required to the plant control and safety systems.

5.   The existing safety and licensing envelopes are maintained by the use of Low Void Reactivity Fuel (53), which compensates for the shortening of the neutron lifetime and the decrease in the delayed neutron fraction. A systematic review of each of the design basis accidents confirms that the reactor will be licensable when operating with a full core of MOX fuel.

Table 17 **Data relevant to stability and kinetics**

	MOX	Natural uranium
Delayed neutron fraction	0.00383	0.00582
Prompt neutron lifetime	0.00050 s	0.0009 s
Fuel temperature co-efficient	-3.0 $\mu k/^{o}C$	-6.0 $\mu k/^{o}C$
Full core void reactivity	-2.8 mk	+11 mk

As the discharge burn-ups of the MOX and natural uranium fuel are similar, no changes are required to the current spent fuel handling and storage methods.

Safeguards and security for the MOX fuel fabrication plant have been examined together with the safeguards and security for transportation of the new fuel to the reactor site. In this regard, input was gained from the International Atomic Energy Agency (IAEA), and the Atomic Energy Control Board (AECB) which is the Canadian regulatory authority.

A comprehensive evaluation of the manufacture of CANDU-MOX fuel has been carried out. This evaluation included participation of a CANDU fuel manufacturer, MOX fuel fabrication experts and AECL fuel design and fabrication experts.

Environmental, Safety and Health issues at the fuel fabrication facility including worker exposure and waste arisings have been evaluated. The fuel transportation system has been designed to reduce radiation exposure to personnel to equal to or less than the current levels with natural uranium fuel. The same packaging system also limits exposure of personnel at the reactor site to less than current experience.

The only changes required to the existing station are associated with storage and handling of the new fuel. A new secure building is required inside the existing PIDAS fence. The route to the new fuel loading room will require hardening. Storage of spent fuel in on-site pools followed by dry storage and eventual burial in a repository was evaluated and confirmed adequate using existing systems designs.

An economic evaluation of the MOX-burning CANDU plant shows that the cost differential between operating with natural uranium fuel and MOX fuel is not of concern. There is no derating of the reactor power required. There are no issues that would lead to a reduction of the current station capacity factor.

Burning plutonium in fast reactors

The CAPRA project

The aim of the first two-year phase of the CAPRA project studies (1993-1994) has been to demonstrate the feasibility of a fast reactor whose net burning of plutonium would be as high as possible and which would, moreover, contribute to the destruction of minor actinides. The greater part of the effort focused on a thorough study of the reference option constituted by cores using a high plutonium content oxide fuel. The potential of an alternative option involving a nitride fuel was also assessed. An exploratory investigation was made on a study of uranium-free cores allowing the highest plutonium burning rates, *i.e.* about 110 kg/TWh (see Figure 22 which gives the net plutonium consumption per TWh as a function of the plutonium content in the fuel).

It is essential to stress that the work of the project concentrated on maximum admissible plutonium content for the oxide fuel, degraded isotopic quality of the reference plutonium (recycled twice in a MOX-PWR), high power core (1 500 MWe), while seeking maximal compatibility with the main options of the EFR nuclear steam supply system and with technological experience regarding fuel.

This work has been performed in the framework of the European R&D collaboration on EFR and in close co-operation with the EFR associates. The R&D programme in support of this activity is complemented by various international collaborations with Japan, Russia, Switzerland and Italy.

Parametric studies indicated that a considerable reduction of the fuel inventory (or "dilution") is always necessary, resulting in a decrease in in-pile fuel residence time, as well as in a reduction

110

(favourable) of the sodium void reactivity, whereas a decrease of the uranium content of the fuel brings about a reduction of the Doppler effect, a decrease of the conversion ratio which causes a large daily reactivity loss.

Figure 22. **Net plutonium consumption as a function of the plutonium content of the fuel**

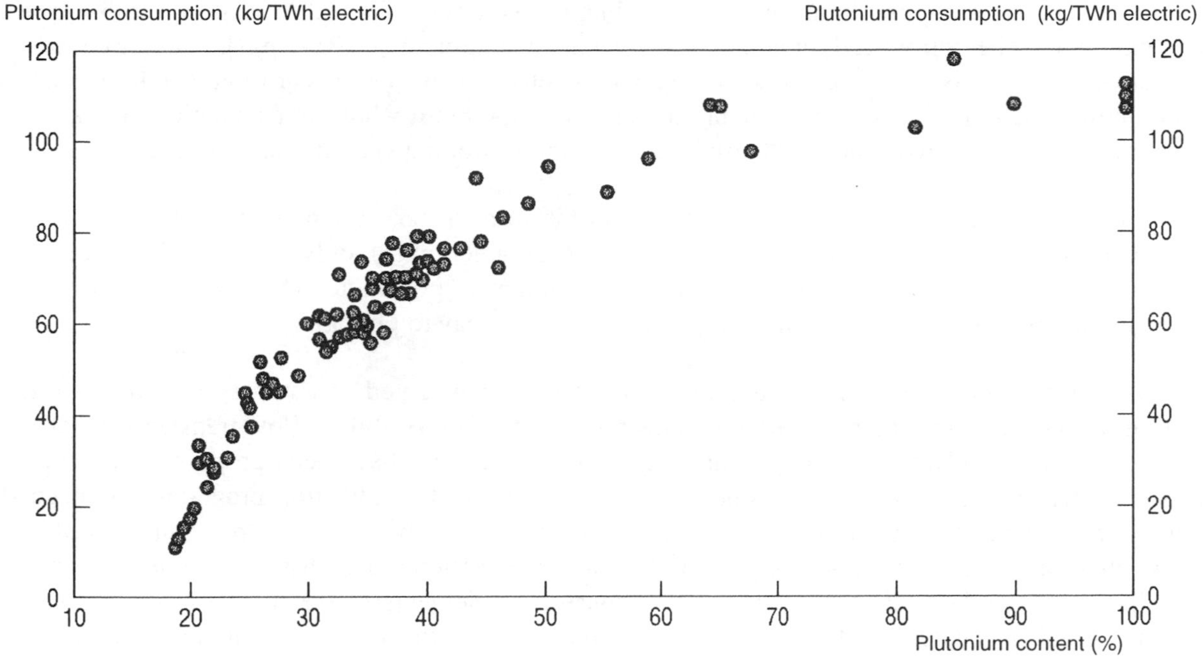

It was demonstrated that cautiously resorting to poisoning, in which the excess reactivity linked to the increase of the fuel plutonium content is compensated by the introduction of a neutron absorber, allowed reducing the drawbacks involved in the dilution process (short in-pile fuel residence time, for example) even if at the expense of a deterioration of the sodium void and Doppler effects. In turn, the use of moderating materials can improve the situation.

The management of the marked reactivity loss and the optimisation of the relative values of the Doppler and sodium void effects appear to be the two major issues to be dealt with so as to ensure the feasibility of plutonium burner cores. In this respect, heterogeneous core designs, in which a part of the inert material replacing the fuel in the dilution process is gathered in specific sub-assemblies, are of particular interest.

The search for compatibility with current fuel cycle technology (dry route fabrication, Purex process reprocessing) led to the choice of a maximal plutonium content of 45 per cent in the oxide fuel, corresponding to a plutonium burning rate of the order of 70 kg/TWh.

To implement 45 per cent plutonium oxide fuel in a 1 500 MWe core implies the use of pins of small diameter containing oxide pellets with a large central hole, by heterogeneous sub-assemblies comprising a large number of pins (469), about one third of which would be empty of fuel but filled with an inert material (the best candidate appears to be spinel, $MgAl_2O_4$), and by the presence of around fifty diluent sub-assemblies containing no fuel. These diluent sub-assemblies, which constitute an essential particularity of the CAPRA design, were accommodated in a core whose volume was that of the EFR reference core by designing a sub-assembly with reduced pitch.

111

An irradiation programme designed to validate and optimise these main fuel choices was defined and launched in 1994.

Regarding minor actinide destruction, the main recommendations to-date obtained from the SPIN programme (see section 3.3) are applied to the reference CAPRA core: homogeneous recycling of neptunium and heterogeneous recycling of americium.

Studies concerning the homogeneous recycling of neptunium showed that up to 3 to 5 per cent neptunium can be introduced into the oxide, while maintaining a (Pu+Np)/(U+Pu+Np) ratio of 45 per cent, since, vis-à-vis reprocessing, neptunium behaviour would appear to be similar to that of plutonium. The core thus contains about one tonne of neptunium whose transmutation rate is about 50 per cent for the approximately 900 efpd (effective full power days) total fuel irradiation.

The heterogeneous recycling of americium in the form of target sub-assemblies arranged in a first peripheral ring around the core (about 1.5 t of americium) only slightly alters the core performance. About 75 per cent of the initial americium is transmuted, 50 per cent of which is changed into plutonium and short-lived curium which will decay to plutonium.

As far as plutonium without uranium fuel options are concerned, the studies carried out in this field have indicated that there are no problems regarding safety (stability, flow transients, reactivity transients, core accidents). Some promising generic fuel families have been proposed, but no clear and definitive choice has emerged, since a specific experimental validation programme is needed. However, the requirement is to find a fuel solution in which any power rise would instantaneously put either the expansion or the Doppler reactivity effect into action. This assumes excellent thermal coupling between the plutonium and the other constituent elements of the fuel. For the expansion effect, moreover, the design of the fuel must allow the latter to expand freely within its cladding.

In summary, the feasibility of a plutonium-burner reactor based on the use of oxide fuel has been demonstrated. Its limits in terms of plutonium burning are 70 to 80 kg/TWh corresponding to the choice of a maximum fuel plutonium content of 45 per cent. The feasibility of reactors based on the use of uranium-free fuel has not yet been fully established. The exploratory studies that have been performed up to now have shown that there is, a priori, no obvious argument as to infeasibility, and that this path should be investigated further.

Burning plutonium in LWRs

A concept of high conversion Pressurised-Water Reactor (PWR) has been developed in JAERI since 1985. In order to increase fuel utilisation or conversion ratio, the reactor would have a triangular pitched tight lattice core with lower moderator-to-pellet volume ratio (Vm/Vp) than that of an ordinary PWR. Since in recent years the function of in-core plutonium storage has been recognised to be more important than the high conversion ratio, a concept of an active storage PWR was derived from the design study of a high conversion PWR. In the active storage PWR core, a large amount of plutonium is loaded and 20 to 25 per cent of the plutonium loaded annually would be consumed.

The R&D status in JAERI is as follows: a) the core design has been completed and the optional value of Vm/Vp was chosen to be 1.33; b) small scale DNB experiments using 7-rod test sections were performed and a prediction method using a CHF correlation and a sub-channel analysis code was evaluated. The experiments using these facilities were completed in 1993. A pressure vessel hydro-dynamic mock-up test facility was also scheduled to be operated in 1996.

A preliminary neutronic calculation indicated that changing the control rod arrangement allows the square lattice core of current PWRs to be replaced by plutonium fuel assemblies. The number of control rods in the core of the active storage PWR should be larger than that in an ordinary PWR in order to compensate excess reactivity.

The plutonium core of an active storage PWR can be replaced by uranium core if fast breeder reactors are commercially introduced in the next century. Therefore, this reactor concept can cope with any future conditions that society might apply to plutonium supply and demand.

3.3   Advanced concepts

Introduction

There is no adopted strategy in the world regarding fast reactor development, but until now only Liquid Metal-Cooled Fast Breeder Reactors (LMFBRs) have received sufficient funding allowing for building and operating prototype breeder reactors (Phénix, PFR, BN-600 and Monju) and constructing a prototype commercial breeder reactor (Superphénix). The foreseen availability of uranium has reduced the incentive for breeding and so a Liquid Metal Fast Reactor (LMFR) which is a net consumer of plutonium is currently being considered to improve the long-term management of both plutonium and TRU waste arising from thermal reactor programmes.

Although considerable experience is already available throughout the world in producing MOX fuel for LWRs, as well as for LMFBRs, it is a challenge to replace oxide fuels by other materials for a future LMFR system. Both plutonium-uranium nitride and carbide, as well as metallic fuels, have been studied as advanced fuels because of their relatively high thermal conductivities and high fissile densities. These characteristics may offer enhanced performance: high heat rating, low reactivity swing and high burn-up. The use of nitride is subject to a prior ^{15}N enrichment process, otherwise large inventories of ^{14}C will be generated during irradiation. The formation of solid solutions of actinides allows wider selection for fuel design compared to the limited plutonium content in the case of MOX and metallic fuels. Carbide is less advantageous because of its very poor handling ability due to its chemical instability. A pilot scale experiment for the production of metal fuel has taken place in the US, while carbide and nitride fuels have only been prepared on a laboratory scale; for example at the ITU of the European Commission, CEA Cadarache of France, PSI of Switzerland and JAERI of Japan.

In the far future new reactor concepts like molten-salt reactors might become more attractive for plutonium and MA burning.

Partitioning and transmutation (P&T) of radioactive waste nuclides is technically feasible, but extensive R&D remains to be done before it can be regarded as a mature option. The incentive for P&T stems from the interest in reducing the long-term hazard of high-level radioactive waste, notably from long-lived actinide isotopes. The effectiveness of the transmutation process is strongly dependant on the neutron spectrum. The concepts under review are based on fast reactors or high flux particle accelerators of various types.

The LMFRs are capable of reducing the plutonium and MA inventories, but the time-scale necessary to implement such systems is rather long and will definitely require the development of an advanced fuel cycle industry. In the near future this would need to include advanced aqueous reprocessing plants with actinide separation from fission products and fuel fabrication facilities with either homogeneous or heterogeneous actinide recycling. In order to reach a time-scale of, for example, 70 to 80 years for an actinide inventory reduction by a factor of 5, burn-ups of 20 to 25 atom per cent will be required. Such high burn-ups will increase the radiation damage on the organic solvents used in the liquid extraction process and will increase the secondary waste production.

Regarding the CANDU systems, a Canadian study has been performed, according to which, plutonium is annihilated, optionally with the minor actinides, in the absence of ^{238}U. The low fissile requirements of the CANDU reactor, together with the flexibility provided by the on-power refuelling system, result in annihilation in excess of 80 per cent of the initial fissile inventory following a single pass through the CANDU fuel channel. An annual annihilation rate of 1.3 t is thus achieved per GWe of installed CANDU capacity.

Studies on plutonium fuels other than oxide

Recycling metallic plutonium in fast reactors

Research and development on metallic fuel cycle technology is being carried out in Japan at the Central Research Institute of the Electric Power Industry (CRIEPI) as one of the long-term nuclear options. This technology was originally developed by Argonne National Laboratory (ANL) in the USA under the Integral Fast Reactor (IFR) programme. R&D in Japan is focused on evaluating the feasibility of the IFR concept. The studies include:

- irradiation of U-Pu-Zr alloy metallic fuel in LMF(B)R;

- fabrication of metallic fuel rods by injection casting; and

- reprocessing of spent fuel by the pyrochemical process (electro-refining of uranium and plutonium in molten-salt systems using solid and liquid metal cathodes).

From the point of view of non-proliferation, as well as waste management, metallic fuel has the following features:

- Plutonium is always co-deposited with uranium, minor actinides and rare-earth elements in the pyrochemical process. Pure plutonium is never produced.

- The plutonium product contains rare-earth fission products which are relatively highly radioactive. Hence, it has to be remotely handled.

- The plutonium product contains minor actinides, which are recycled within the fuel cycle and burnt in the reactor.

Fuel cycle of nitride fuel for FRs

Nitride fuel is usually fabricated from oxide by a one-step carbothermic reduction in a similar manner as carbide fuels. Oxide with excess carbon is converted to nitride by heating in a stream of N_2-H_2 mixed gas. During conversion, residual excess carbon may be removed continuously after the formation of oxygen-free nitride. Nitride is chemically rather stable so that fuel in pellet-form could be handled even in air without significant oxidation. In addition, a sol-gel method has been investigated for decreasing radioactive dust and enabling an automatic fabrication process. The technology for characterisation such as the determination of nitrogen, carbon and oxygen was also established.

Nitride fuel shows low fission product gas release and no chemical interaction with cladding materials under irradiation. The application of the "cold fuel concept" for nitride fuels may result in a great margin against fuel failure. The maximum burn-up reached 15 to 20 per cent fissions per initial metal atoms (FIMA) in previous irradiation tests in the USA, ex-USSR and European countries. It was also confirmed, by recent irradiation tests of thermally stabilised fuel pellets in JMTR, that nitride fuel reaches burn-ups up to 5 per cent FIMA without failure. The Joyo irradiation test of mixed nitride fuel pins fabricated by JAERI was scheduled to start in August 1994 under collaboration with PNC.

Nitride fuel may be treated by a Purex method applied to MOX fuel. However, an innovative pyrochemical reprocessing method developed for the IFR programme, may also be available for nitride fuel. The concept of a self-completed nitride cycle proposed by JAERI includes pyroprocessing with fused LiCl-KCl salts, electro-refining and fabrication process as for sphere-pack fuel pins, as shown in Figure 23. In the electro-refiner, actinides in nitride may be ionised at the anode together with the formation of nitrogen gas, where actinide metals are deposited at the cathode. The validity of the process was checked by a thermochemical assessment using uranium nitride. The experiment of electro-refining actinide nitrides, such as PuN, is being set up in JAERI.

Helium-cooled fast reactor with nitride fuel

A helium-cooled fast reactor with coated particle nitride fuel has been proposed as a dedicated reactor for actinide burning by JAERI. An actinide kernel with a composition of (66NpAmCm-34Pu)N is coated with thin TiN layers to form a tiny fuel particle having a diameter of about 1.5 mm. The fuel element design uses two concentric porous frits between which the fuel particles are packed. The fuel particles are directly cooled by radial helium gas flow through the particle bed. Since a large heat transfer surface per unit volume of the particle bed is quite effective for heat removal, a very high power density and thus very high transmutation rates are obtained.

A conceptual design study of the reactor is under way. The study includes the neutronic and thermal-hydraulic calculations and processes and flow-sheet studies for the fuel cycle. The thermal power of the reference reactor design is 1 200 MWt and the burn-up is 17.3 atom per cent per year.

Another concept of HTGR which would burn plutonium to the point of almost not requiring reprocessing is also being studied. Pebble bed type HTGRs using plutonium balls (to burn plutonium) and fertile balls (thorium or ^{238}U to breed) can offer almost 100 per cent plutonium burning, because a continuous supply of fresh plutonium balls would compensate reactivity partially lost to maintain reactor operation. Both the plutonium balls and the fertile balls are loaded into the core randomly and they are exhausted continuously from the core. The burn-up of plutonium balls can be evaluated by measuring the strength of γ rays, and if it is still under the maximum then the ball is reloaded again.

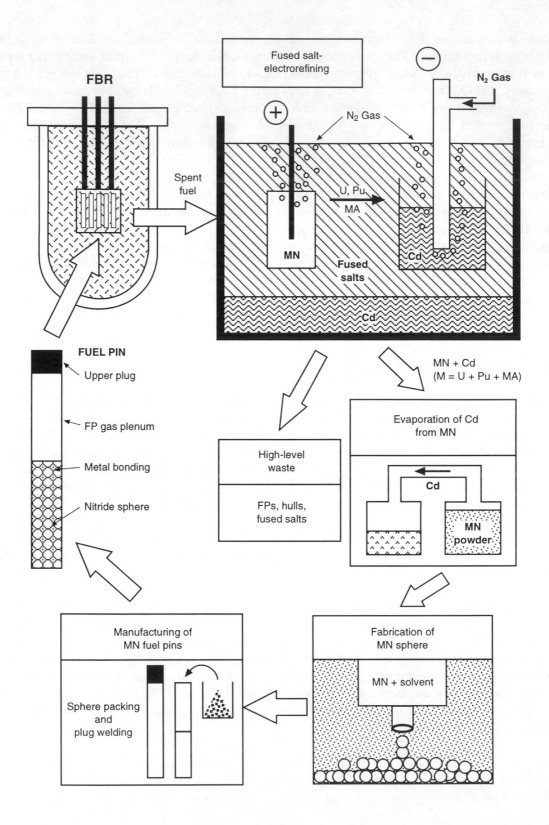

Figure 23. **Cycle coupled with nitride fuel and pyroprocessing**

Burning plutonium and MA in FRs

The SPIN programme

In response to public concern about nuclear wastes and particularly the long-lived high-level ones, a French law issued on 30 December 1991 identified the major objectives of research for the next fifteen years, before a new debate and possibly a decision by the French Parliament on final disposal. These objectives are:

1. improved conditioning of the waste;

2. extraction and incineration of long-lived waste in order to minimise their long-term toxicity;

3. research performed in underground laboratories to characterise the capability of structures to confine radioactive wastes (two sites have to be selected for these underground laboratories, in consultation with the local population); and

4. the study of conditioning and prolonged surface storage of waste.

To comply with the requirements of the December 1991 law related to the management of long-lived high-level waste, the CEA launched an important long-term R&D programme. A part of this programme, called SPIN, is devoted to separation and incineration of such waste.

Plutonium management represents the major issue in order to reduce significantly the radiotoxicity source term of a fuel irradiated, for example, in a PWR. If plutonium is extracted and re-used, the present studies indicate that americium isotopes and, to a lesser extent ^{237}Np, can account for a further significant reduction of the radiotoxicity source term. With an appropriate out-of-pile management of curium, their transmutation can be envisaged by multiple recycling in fast neutron reactors.

Taking into account the specific characteristics of the different elements present in waste and their effects on fuel cycle operations (doses delivered during fabrication, transportation and reprocessing, decay heat), one is led to selective recycling of irradiation products, with limitations in MA content also set by modifications in core safety parameters.

Preliminary studies of an advanced burner reactor simultaneously consuming plutonium and neptunium homogeneously in the core and heterogeneously in the radial blanket, show a balance between the overall production and destruction of these elements in PWRs (both UO_2 and MOX-fuelled) and FRs (FRs participating to about 20 per cent of the global electricity production). The associated radiotoxicity reduction is some 40 times with respect to an open cycle.

In summary, concerning long-term options and guidelines, in France, the following two scenarios can be envisaged:

Scenario 1: Uses the potential of existing PWRs for single plutonium recycling. After reprocessing, the plutonium is transferred to a fast burner reactor, which can also incinerate MAs from both the PWR-MOX and the PWR initially loaded with UO_2.

In addition to its ability to incinerate poor-grade plutonium, the burner reactor generates only small amounts of ^{243}Am and curium. Successive recycling in fast neutron reactors does not damage

117

the plutonium grade; moreover, with appropriate dilution, it can bring it back to a favourable composition for use in any reactor. The number of recycles remains to be determined, depending on other factors, especially on the observed efficiency of minor actinide incineration.

This scenario involves building a number of burners, whose proportion in relation to PWRs will be determined by their technical performance. The ratio could be around one fast reactor to five PWRs. The optimisation of next generation PWR cores for single plutonium recycling should be considered for greater efficiency in plutonium burning, which will limit the number of burners, and for limiting the amount of americium and curium build-up.

> *Scenario 2*:   Uses the same basis as Scenario 1, but optimises the number of plutonium recycles in PWRs to take full advantage of existing reactor facilities and to build a minimum number of burners. In economic terms, the cost of these reactors is assumed to be greater than that of PWRs at present.

Increasing the moderating ratio of PWRs, which results in higher plutonium consumption and lower actinide production, could bring some improvements. The choice of an optimum MOX burn-up will be another key factor in the optimisation sought.

In addition, the number of times plutonium is recycled in PWRs will be limited by its progressive degradation. The goal will still be to ensure good plutonium burning efficiency and to make easier the minor actinide incineration in order to decrease, by an important factor, the total amount of actinides in the waste.

The Japanese programme

The Japanese Atomic Energy Commission has recently published a *"Long-Term Programme on Nuclear Research, Development and Utilisation"*, in which it promotes a long-term R&D activity on advanced recycling technologies. The advanced recycling includes technologies for recycling MA and for using metallic fuel and/or nitride fuel, as well as oxide fuel. The R&D programme aims to improve F(B)R fuel cycle systems by further advancing their safety, reliability and economics, and also aims to reduce their environmental impact while maintaining the Japanese commitments towards nuclear non-proliferation.

The MA mass balance is studied by PNC on the assumption that plutonium and MAs are recovered from LWRs and plutonium-fuelled thermal reactors and repeatedly recycled into fast reactors. It is assumed that nuclear power generation is increased up to 1 000 MWe to 1 500 MWe per year and fast reactors start to be used commercially in the year 2030. New reactors are then assumed to be only FBRs, and LWR reprocessing is assumed to be applied to all spent fuel discharged from LWRs and plutonium-fuelled thermal reactors.

An 1 000 MWe of annual power generation by LWRs, plutonium-fuelled thermal LWRs and FBRs produces an amount of MA, which is transferred into the high-level waste, of a total of 310 tonnes without MA recycling by the year 2100. In the case of MA recycling in commercial FBRs after the year 2030, the MA amount existing in the fuel cycle in the year 2100 is reduced to about 60 tonnes, a reduction of 80 per cent compared to the non-recycling case. The residual accumulated MAs are retained in the fuel cycle. According to this analysis, the maximum total MA accumulation will be reached in about the year 2065. The reduction ratio of curium is relatively low compared with those of americium and neptunium due to the additional generation of curium by the neutron capture of americium in the FBR core.

Plutonium annihilation and immobilisation

CANDU

The CANDU reactor can annihilate plutonium together with the minor transuranic actinides in the absence of ^{238}U. An annihilation rate of over 80 per cent is achievable per pass through a fuel channel. Use of a second or a third pass eliminates the need for reprocessing and refabrication cycles.

This ability of CANDU is related to the basic design features of using heavy water as moderator and a fuel handling system that provides an on-power refuelling capability. Both design features lead to low neutron absorption rates in the CANDU lattice due to the low neutron absorption cross section of the heavy water and the absence of burnable poisons for reactivity suppression. As a result, CANDU can be operated with a variety of low-grade (low fissile content) fuels as these provide sufficient reactivity in a CANDU lattice.

For plutonium annihilation in current CANDU reactors, the fuel consists of a mixture of the plutonium isotopes in a neutronically inert matrix. (Several materials for the matrix are being developed.) The absence of ^{238}U eliminates the main source of the plutonium isotopes. It also eliminates the main neutron absorber of the lattice resulting in a remarkable improvement in neutron economy. The loss of the neutron absorption in ^{238}U, which comprises over 30 per cent of the neutron absorption in the CANDU lattice, reduces the fissile requirement of the CANDU AB (Actinide Burner) to a fraction of the fissile content of natural uranium.

A lower fissile inventory requires a correspondingly higher operating neutron flux level to produce the rated power. The annihilation process in CANDU involves, first, neutron capture to change from even to odd and, then, fission by thermal neutrons of the odd-numbered isotopes. The high thermal neutron flux level provides an appreciable annihilation rate in spite of the neutron capture required first. In particular, the on-power refuelling system is used to move the fuel into regions of higher flux as its fissile content depletes during irradiation. Through proper fuel management of this type the last traces of plutonium can be eliminated before the fuel is discharged.

The absence of ^{238}U has a major impact on the fuel management strategy that is used in plutonium annihilation. As formation of ^{239}Pu is eliminated, the reactivity of the lattice drops rapidly with fuel burn-up. The refuelling rate required to maintain criticality is significantly higher compared with the reactor that burns natural uranium. The refuelling rate is kept within the capability of the Fuel Handling System by adjustment of the initial fissile content of the bundles. An increase in fissile content, however, requires reactivity suppression which reduces the thermal neutron flux level and subsequently increases the plutonium content of the discharged bundles.

The absence of ^{238}U essentially eliminates the coolant void reactivity effect as void reactivity in CANDU is a result of the behaviour of the neutron cross section of ^{238}U on coolant voiding. The fuel temperature coefficient, however, remains essentially the same as for natural uranium (close to zero or negative). This is especially the case with plutonium having a high ^{239}Pu content as the value of the ^{239}Pu η in a CANDU lattice decreases with fuel temperature increase.

Concerning plutonium use, basic research has been performed aiming at almost completely burning plutonium in LWRs, making it possible to dispose of it directly. This process consists of producing chemically stable (rock-like) fuels in conventional fuel facilities, burning such fuels in LWRs and disposing of them without further processing.

Concepts relate to certain rock-like fuels, which are tailor made multi-phase fuels consisting of mineral-like compounds, that are, chemically and thermodynamically, so stable that they are not soluble in nitric acid in normal ways. During irradiation in reactors, most of the solid fission products would be solidified by their substitution in the matrix compounds and/or precipitated as new mineral-like compounds in the fuels. The phases thus obtained in the spent fuels would become stable for geological periods. Accordingly, they would be suitable for direct disposal without further processing.

Both from the discussion and the experimental results on phase relations, chemical and geological properties of minerals and ceramic materials, it is likely that thoria (ThO_2) and stabilized zirconia (ZrO_2) with fluorite type structure, as the host phase for PuO_2, are suitable additives to dissolve actinide and lanthanide elements. For the chemical stability of the spent fuels, alumina (Al_2O_3) and magnesia (MgO) would be suitable additives. Of these, both fuel systems PuO_2-ThO_2-Al_2O_3(-MgO) and PuO_2-ZrO_2-Al_2O_3(-MgO) seem to be favourable. As indicated from reactor burn-up calculations, conventional LWRs are considered to be suitable for their burning: about 98 per cent of ^{239}Pu (*i.e.* 85 per cent of total plutonium) can be transmuted in initially loaded high-grade plutonium containing approximately 94 per cent ^{239}Pu. The spent fuels thus obtained would have high chemical stabilities, high neutron emission rates, high heating rates, high radioactivity and poor ^{239}Pu quality. These make the spent fuel unattractive for subsequent use.

3.4   Geological disposal of separated plutonium

Introduction

A thorough evaluation of the issues associated with the management and disposition of plutonium, concentrating primarily on weapons plutonium, but also considering civil plutonium, was completed in 1994 by the Committee on International Security and Arms Control of the US National Academy of Sciences (NAS) (9). The committee concluded that there are a number of feasible disposition options, and that these methods should be so designed to place the excess plutonium in a physical form that is at least as inaccessible for future weapons use as the plutonium in spent fuel from civil nuclear reactors. This premise will be adopted in the following discussion of permanent disposal options. The NAS Committee determined that the two most promising alternatives for achieving long-term plutonium disposition are its fabrication as mixed oxide (MOX) fuel and use in existing or modified nuclear reactors, or its vitrification in combination with high-level radioactive waste for ultimate disposal in a geologic repository. The committee also addressed a third option, direct disposal in deep boreholes. After the completion of the work of the NEA expert group, the USDOE published a Technical Summary Report For Surplus Weapons-Usable Plutonium Disposition (11).

This section deals with the options for permanent disposition of civil plutonium in geological structures. Other sections in this report treat interim storage of plutonium and its use in current-generation light-water reactors or future advanced reactors. Disposal of spent fuel generated by reactor options is not considered in any detail here, because that topic has been adequately reported in previous publications.

Disposal options

The disposal options considered below include the two non-reactor options, vitrification and burial in deep boreholes, as recommended in the NAS study. Also treated here are variants of the vitrification option that have recently been identified as having promise: both involve the immobilisation of plutonium in ceramic or metallic forms together with fission products. The vitrification option and its variants would all yield a product that is self-protecting and resistant to retrieval of the plutonium, rendering it virtually as inaccessible as in spent nuclear fuel. The ultimate waste form would be disposed in a mined geologic repository. Other options such as sub-seabed disposal, underground detonation, deep space disposal, or ocean dilution were discounted in the NAS study and will not be covered here.

Geologic repository disposal

For the purposes of this discussion, geologic repository disposal will be defined as disposal in a deep underground mined structure that has as its primary purpose the permanent disposal of high-level radioactive waste, including intact spent nuclear fuel elements. In order to place the separated plutonium in such a structure and preserve the attributes of highly restricted accessibility, the plutonium must be introduced into a processing system that changes its physical or chemical form, thus permitting its incorporation with other materials that confer the desired attributes. This processing of this plutonium can be simple or highly complex, depending upon the avenue chosen for its disposal form and repository characteristics.

Preparation for disposal by vitrification

A number of processes and equipment are in commercial use throughout the world that make borosilicate glass for a variety of large and small applications. Most of these could likely be adapted to produce plutonium borosilicate glass. Although there are many differences among these applications, including feed composition, melter design and heating method, the compositions and properties of most of the final glass products are similar. Heating methods that are in use include joule heating (heating by resistance to the flow of electricity), induction, microwave, plasma (high-temperature gases) and radiant energy. All of these heating methods work well, and the choice depends on the particular application, feed composition, throughput capacity, and final form criteria. Vitrification of high-level radioactive waste is now becoming more prevalent, worldwide.

For vitrification of high-level radioactive waste, the oldest and most thoroughly demonstrated process is the one that was used at Marcoule in France, which, with minor alterations, is being used in the United Kingdom and France (La Hague). This process calcines acidic waste, feeding the dry calcine to the melter. In the US, most of the high-level radioactive waste (Savannah River Site and Hanford) are alkaline, and contain significant quantities of sodium and aluminum. For vitrification of these wastes it was appropriate to use a liquid slurry-fed melter. A similar melter process is used in Belgium and Japan. The US melter has joule and radiant heating, and a cooling water jacket that freezes the glass in the interior insulation (preventing molten glass from reaching the melter's steel

121

shell) and maintains the exterior at a low temperature to prevent up-drafts. To satisfy waste form criteria, the melter design provides for sampling of the molten glass as it pours into the canister.

Borosilicate glass vitrification of plutonium appears to be a technically viable option. The technical basis for the method derives from over 30 years of development and application of borosilicate glass stabilization of high-level radioactive waste from reprocessing of spent fuels and targets. This stabilization method has gained wide international acceptance. High-level radioactive waste vitrification plants are in operation in France, the United Kingdom, Belgium (in a joint project with Germany), and Japan. A plant is now operating in the United States, and an additional plant in Japan is under construction. During the development of these technologies sufficient data were obtained on the behaviour of uranium and plutonium in this type of glass to give some confidence that plutonium can be incorporated in the glass.

The large-scale plants in Europe and Japan that have vitrified high-level waste have demonstrated that plutonium that is present in incidental very small concentrations in the high-level waste can be successfully incorporated in the glass. In addition, there are numerous reports of laboratory experiments in which there has been incorporation of several per cent plutonium by weight in such glasses, with and without simulated high-level waste constituents being present. Laboratory tests at CEA (France) and KfK (Germany) have produced simulated waste glasses with 3.5 to 4.5 wt. per cent plutonium. Large-scale vitrification of high-level waste using an aluminophosphate glass was initiated in Russia in 1991. The joule melter is still in operation. The plutonium concentration in the glass is less than 0.1 wt. per cent. Laboratory experiments there showed that the maximum concentration of plutonium in this type of glass is in the range 1.0 to 1.5 wt. per cent, and rapidly decreases with the increase of aluminum concentration in the glass matrix. Laboratory work is also underway in the United States at the Savannah River Site to develop plutonium glasses without high-level waste components. Some laboratory experiments have produced experimental glasses with 7 wt. per cent to 15 wt. per cent plutonium, but these glasses have not yet been fully characterised, so that their performance over long times in a repository is not known. This type of plutonium glass, *i.e.* without high-level waste components, could be used for interim storage of plutonium; encapsulated plutonium in this form would be critically safe, stable and monitorable. Later, the plutonium could be retrieved, if needed for energy production, or it could become feed for a final vitrification step to produce a glass containing 1 to 4 wt. per cent plutonium and fission products that would be placed in a geologic repository. Vitrification without added fission products could be performed in shielded glovebox facilities. However, incorporating high-level waste or radioactive spikes into the glass would require heavy shielding, remote operations, and large cell or canyon-like facilities.

Physical security and nuclear criticality avoidance are both significant issues if plutonium borosilicate glass is to be placed in a repository. It is likely that the international community would desire international safeguards of the plutonium for a very long time, which adds to the cost. The potential for criticality from either spent fuels or plutonium borosilicate glass in a repository over geologic time is an issue which is now being seriously assessed by the technical community.

The chemistry of the plutonium glass-making process requires further laboratory development and demonstration for this application. Tests must determine the optimum composition of the glass for this specific purpose. Properties of the glass product, such as durability and plutonium leach rate, must be determined. Preliminary calculations (54) indicate that the boron and lithium in the Savannah River Site's reference glass will assure criticality safety up to about 14 wt. per cent plutonium (weapons-grade) as long as the neutron absorbers and the plutonium remain together. Other, more detailed, calculations must be made that include other glass compositions, other absorbers (e.g. erbium or gadolinium), a variety of neutron moderating conditions, fast neutron conditions, and other plutonium isotopic compositions.

The leach rates of plutonium, boron and lithium from glass are reasonably well known; boron and lithium will leach out much faster than plutonium and it is assumed that in a repository there is a possibility over geologic time for separation of these neutron absorbers from the plutonium, which raises the issue of nuclear criticality. The effects of other absorbers on plutonium solubility and glass properties must be determined, as well as the leach rates of other candidate poisons. The volatility and entrainment of plutonium into the melter offgas must be determined as a function of melter temperature, offgas flow rate, and other process and equipment variables. Any measurable carry-over of plutonium into the offgas will require careful design to prevent significant accumulations of plutonium in the offgas system that could lead to nuclear criticality accidents. Design features must also assure that criticality cannot occur in the melter feed tanks or in the melter itself.

Immobilisation of plutonium in a ceramic waste form

In addition to borosilicate glass, there are a number of other waste forms promising acceptable performance. Synroc (synthetic rock), a ceramic form developed in Australia, has emerged as a potential second-generation waste form, but it has not yet been commercialised; the process technology has been demonstrated at commercial scale with simulated high-level waste. An extensive data-base has been generated on the aqueous durability of Synroc on laboratory-scale specimens containing radioactive high-level waste and high loadings of $^{244}Cm_2O_3$ (4 wt. per cent) and $^{238}PuO_2$ (11 wt. per cent). Strategies for immobilising plutonium, with or without added fission products, and for incorporation of neutron poisons for criticality control during processing and disposal have been developed (55).

The basis for the ceramic immobilisation form is the incorporation in solid solution of the fissile material within the structures of ceramic phases that are identical to leach resistant minerals that have demonstrated an ability to immobilise natural radioactivity over geological time scales. There are many natural minerals that can accept some actinides and fission products into their crystalline structures. However, a multi-phase assemblage of the leach resistant minerals is necessary to incorporate and immobilise the full set of fission products and actinides in high-level waste at practical levels of waste loadings. Thus, Synroc consists mainly of the titanate minerals, zirconolite $(CaZrTi_2O_7)$, barium hollandite $[Ba(Al,Ti)_2Ti_6O_{16}]$, perovskite $(CaTiO_3)$ and excess titanium oxides; hollandite acts as the host for Cs, Ba and Rb, perovskite is the major host for Sr and zirconolite and perovskite are the hosts for lanthanides and actinides. The composition of Synroc can be varied, depending on the types of materials to be immobilised.

For immobilising plutonium only, a zirconolite-rich variant of Synroc is eminently feasible (56) and, for additional protection against diversion, the hollandite can incorporate approximately 1 wt. per cent (total) of ^{137}Cs. If it is desired to incorporate reprocessing high-level waste to provide diversion resistance, the proportion of zirconolite can be reduced as necessary. The plutonium loading in Synroc is in the range of 10 to 30 wt. per cent and rare earths, hafnium or other neutron poisons can be added also.

Electrometallurgical treatment

Another approach for the disposition of separated civil plutonium is the physical combination of plutonium with the radioactive constituents of spent nuclear fuel. The advantage of this approach is that it achieves spent fuel inaccessibility of the plutonium without requiring fuel fabrication and further reactor irradiation. A promising method for denaturing excess plutonium is an electro-metallurgical treatment process that has been developed for the treatment of spent fuel (57, 58). With this method, waste forms can be prepared for geologic disposal having plutonium contents that are limited only by repository criticality considerations. Plutonium contents as high as 10 to 25 wt. per cent are possible, but the higher quantities may not be practical, due to criticality concerns; 10 per cent may be taken as a nominal value.

Electrometallurgical treatment processing technology has been developed in the United States for the disposal of spent nuclear fuels that are not easily qualified for direct repository disposal (57). The process separates pure uranium from the spent fuel and, in some spent fuel types, extracts inert materials for disposal as low-level waste; the alkali metal, alkaline earth and rare earth fission products and the transuranic elements are retained in the electrorefiner electrolyte salt and are immobilised by absorption in a zeolite ion exchange column, whereafter the loaded zeolite is mixed with glass frit and hot-pressed to form a glass-crystalline ceramic waste form. The spent fuel cladding is recovered and melted separately, and serves as a matrix for the immobilisation of the transition metal/noble metal fission products. This same electrometallurgical treatment technology can be used for excess plutonium disposal by placing plutonium metal (if necessary, after reduction from PuO_2) in the electrorefining cell and combining that plutonium with the spent fuel transuranics and fission products in the ceramic waste form. This provides isotopic dilution of the excess plutonium and also confers the desired degree of radioactive self-protection. The process can be operated with plutonium in metal or alloy form, as well as with plutonium oxides and chlorides.

Packaging and deep borehole disposal

The primary alternative to a high-level radioactive waste repository for the ultimate disposal of plutonium is development of a custom geologic facility designed just for plutonium. A variety of geologic facility types could be considered, but the concept currently under serious consideration is the deep borehole which was mentioned in the NAS report as worthy of further consideration (9).

Deep borehole disposal has been considered in recent decades for disposal of both hazardous and radioactive wastes. This concept received significant investigation in the 1970s for disposal of high-level radioactive waste including spent nuclear reactor fuel. Several limitations in the concept for that particular mission led several nations to drop it in favour of a mined geologic facility (59). Examination suggests the reasons for rejecting deep borehole disposal of high-level waste and other wastes are not prohibitive for a plutonium disposal mission, and may even become assets. These issues and counter-arguments for the plutonium disposal mission include the following: a) retrievability of high-level waste from a deep borehole would be difficult (but non-retrievability may be *desired* for plutonium disposal); b) the volume of high-level waste would require many large holes (but the volume of separated plutonium is comparatively small); c) heat generation of high-level waste would limit hole capacity (but the heat generation rate of excess plutonium is comparatively small); d) the level of isolation required for intermediate-level waste does not justify the cost (but there is a high priority on isolation of plutonium to ensure against recovery and misuse of the material); and e) there were limitations on drilling technology (but the technology has improved greatly in recent years).

The primary technical concern unique to plutonium disposition is that of criticality control, which must be considered for any concept placing significant quantities of fissile material in an uncontrolled environment, whether it is a repository or a deep borehole.

Plutonium could potentially be disposed directly by placing it, in metal or oxide form, in small criticality safe containers and assembling these containers into a larger emplacement canister or rack. Criticality control would be designed by physical dispersal, neutron absorbers, or both. Immobilised plutonium could also be emplaced. Suitable forms could include vitrified glass, cement, ceramics or metal alloys. These waste forms could be assembled into larger packages for efficient emplacement. Emplacement techniques developed for underground weapons testing serve as a valuable technical basis for controlled placement of large assemblies underground. It has also been suggested that plutonium loaded materials, such as glass marbles, ceramic pellets or metal alloy beads could be used in or as the aggregate in a stemming grout and pumped into the bottom of the hole and cured in-situ. This results in high volumetric efficiency and complementary use of both dispersion and neutron poisons for criticality control. Capacity per hole could vary from a tenth of a tonne to tens of tonnes of plutonium.

The concept of "spiking" plutonium in a reactor at low burn-up to complicate diversion could result in a form suitable for deep borehole disposal, although this option would incur the added costs of a nuclear fuel cycle and the complications of handling gamma-active materials. The thermal output, high radiation fields and inventory of soluble fission products raise potential problems of safety, operational and regulatory issues.

Perhaps the greatest unknown for the current feasibility of the deep borehole option, or any custom final plutonium disposal, geologic or other, is the uncertain regulatory status. The majority of radioactive waste management regulations around the world evolved without separated plutonium being considered as a waste. Applicable regulations for siting, constructing, operating and closing a plutonium specialised disposal facility do not exist, and would have to be considered in parallel with any serious development project.

Disposal considerations

Impacts of plutonium in geological disposition

Most plutonium disposition options generate one or more high-level radioactive waste streams which are not included in the current plans for the existing high-level waste geological repository programmes of most countries. Disposition of plutonium in a geological repository raises a new set of issues and potential impacts. The severity of these impacts depends on both the nature of the plutonium waste forms considered and on the baseline design of the repository. Waste forms differing from those currently expected have significant impacts on the waste management system because of various unique attributes, as listed below:

1. The inventory of radionuclides and their speciation (metal, oxide, carbide, particulates, organics, etc.) are important in the source term for repository performance analysis. Plutonium disposal forms may have significantly different actinide inventories compared to standard spent nuclear fuel and high-level waste. Even where standard forms have significant actinides (such as spent fuel), the details of the chemistry, such as the oxidation state of plutonium, may differ in plutonium waste forms.

2.   The physical form and dimensions of the plutonium to be disposed may have significant impact on handling, packaging, storage and retrieval. Chemical characteristics are important for compatibility and avoidance of adverse reactions or response to elevated temperature air and water. Mechanical fragility or vulnerability could complicate handling. Examples of forms with such impacts included advanced reactor fuels, accelerator transmutation targets and custom designed immobilisation forms.

3.   Nuclear reactivity becomes an important issue when dealing with significant quantities of plutonium. Criticality can be controlled by physical dispersion, neutron absorption, moderator exclusion and material mass limits. Criticality control must be considered under several regimes: as emplaced, but considering potential external events such as water intrusion, after physical degradation of the waste package and waste form which allows reconfiguration of fissile materials, and allowing for geochemical processes which could reconcentrate actinides and leach neutron absorbers. Issues include: how can criticality control be assured, the impact of adding or removing moderation, physical form degradation, long-term changes in isotope mix due to decay and in-growth, and ultimately the potential consequences should criticality occur.

4.   The radiation emitted from waste forms containing plutonium could be significantly different from similar forms without plutonium. Adding plutonium to a high-level waste glass matrix will lead to generation of neutrons from (α-n) reactions with low-Z materials. Minor actinides, which accompany the plutonium, have significant radiation output. Specific heat output as function of time will also vary for plutonium-bearing wastes. High-level waste with plutonium added will have significant increases in thermal output at long times. Also, thermal properties such as conductivity will vary.

5.   The long-term mobilisation and release of radionuclides is the fundamental measure of repository performance. This will depend on all of the attributes discussed above and their interaction with the repository environment.

6.   Waste forms containing substantial plutonium which could be separated chemically may be potential diversion threats. The standard for comparison is the diversion resistance of spent light-water reactor fuel. Immobilised forms can be expected to contain larger fissile material inventories than spent light-water reactor fuel, and may be easier to chemically process to remove plutonium than standard spent fuel. The issue of potential diversion is central to plutonium disposition options.

The addition of plutonium disposition wastes to the mission of current high-level waste management programmes could result in impacts to existing or planned facilities in several areas, including: a) transportation and interim storage;   b) repository design and operation;   and c) repository performance prediction and licensing.   Facility impacts could include safeguards, radiation and thermal output, required capacity and public perception. Design and operational impacts could result from safeguards requirements, physical or chemical attributes of the wastes and thermal and radiation output.   Waste package designs may require changes to accommodate plutonium wastes and assure criticality control.   Long-term performance predictions and other licensing arguments could be affected by the radionuclide inventories, chemical attributes, leaching characteristics and criticality issues. Licensing could be delayed by adverse public perception.

The least severe impact should result from disposal of forms similar to those expected by the repository. The primary example of this is emplacement of spent MOX fuel in a repository designed

126

for disposal of spent oxide fuel in a once-through fuel cycle. Here the physical characteristics of the MOX fuel are very similar to the repository design basis, with only minor variations in radionuclides, radiation and thermal output. The primary impacts might be in criticality control due to the higher fissile material inventory in typical spent MOX fuel, and in the potential requirements for safeguards on the repository. Another difficult to define impact could be a change in public perception of the facility. On one hand, the plutonium fuel could be perceived as a greater hazard and on the other hand the facility could be perceived as having a greater societal value by virtue of reducing the hazard of excess plutonium. Additional impacts would result from disposal of new reactor fuels designed for plutonium burning in reactors and significantly different fuel types, such as that used in high-temperature gas reactors. These fuel types would have physical, chemical and performance characteristics which would require significant characterisation to assess disposal suitability and performance.

All of the immobilisation forms (glass, other ceramics such as Synroc, or metal alloys) would require assessment for criticality control, radiation effects and long-term performance. Aqueous leaching of radionuclides from plutonium-loaded glass may be different than from glass with just fission products. For repository programmes where vitrified high-level waste is the only planned waste form, addition of plutonium would be the first introduction of actinide effects. The effects of alpha decay on the waste form should be further assessed, as should the longer half-life of the thermal decay curve which would result in somewhat elevated temperatures for much longer times than with just fission products present. In addition, significant quantities of fissile material would result in criticality control and long-term safeguards being added to the list of issues.

In summary, there are a variety of options for plutonium disposition which result in geologic disposition of unique wastes. Suitability and performance of these wastes must be assessed against the particular requirements of the disposal facility for which they are destined.

Criticality safety

Criticality prevention is an absolute requirement for all steps in the disposal of plutonium, even including the period during which the material is in its ultimate disposal location, a geologic repository or deep borehole. In the case of disposal in glass, boron and lithium in the borosilicate glass composition, together with certain fission products (if present), act as neutron absorbers and should serve to limit the effective neutron multiplication factor, k_{eff}, at least over the period of time during which the integrity of the containment system can be assured. Once the containment is breached and water gains access to the glass, the boron can be expected to be removed by leaching at a much faster rate than the plutonium. Therefore, it must be assumed that the beneficial criticality-limiting attributes of the glass will not be available over the lifetime of the primary fissile plutonium isotope, ^{239}Pu, which has a half-life of more than 24 000 years. Furthermore, ^{239}Pu decays to the fissile isotope ^{235}U, which itself has a half-life of 700 million years, but a larger critical mass. The same sort of concerns pertain to the disposal of plutonium via the electrometallurgical method. The behaviour of the waste forms may differ somewhat over time periods of tens of thousands of years, but the comparative performance on a geological time-scale must certainly be indistinguishable.

Physical protection, accountancy and control measures for geologic disposal of plutonium are necessary whether the plutonium is embedded in spent fuel (a per cent or less for LWR spent fuel, a few per cent for MOX spent fuel), or embedded with high-level waste components in glass (a few wt. per cent plutonium), crystalline ceramics (approximately 10 wt. per cent plutonium) or other material such as metal (on the order of 10 wt. per cent plutonium). The addition of radioactive material such as high-level waste or ^{137}Cs to plutonium immobilised in glass or other material has been proposed to make it as inaccessible as plutonium in spent fuel. The measures that must be imposed on geologically-disposed plutonium will, until the repository is permanently sealed, be similar to those required for surface storage of separated plutonium. Inventorying would likely be supplemented by containment and surveillance methods such as portal-perimeter monitoring. Sensors might include motion and seismic sensors on the perimeter and gamma and neutron radiation detectors at the portals. In addition, tamper-indicating or warning sensors could be attached to the containers. Inventorying plutonium disposed of in a deep borehole is not practical.

Materials control and accountancy for plutonium disposal should be similar to that for spent fuel and should not present any unusual difficulties. It amounts to ensuring that all items (fuel assemblies, glass logs, etc.) are accounted for and have not been tampered with. Tags can facilitate inventorying, which could be done remotely, and seals or constant surveillance could ensure that no material has been removed. Inventorying and surveillance of individual containers would begin at the facility where the plutonium is immobilised. Inventorying is more difficult and important prior to immobilisation. However, methods employed to safeguard MOX fuel fabrication facilities and large reprocessing facilities could be employed in a facility that immobilises plutonium mixed with radioactive materials.

The security of disposed plutonium rests on various characteristics (barriers) of the disposal system. The isotopic make-up of the plutonium will not be a major deterrent unless there is a large amount of ^{238}Pu. The ^{238}Pu isotope generates about 560 watts per kilogram and could preclude the fabrication of a weapon. The typical amounts of ^{238}Pu, ^{240}Pu and ^{241}Pu in reactor-grade plutonium complicate, but do not preclude, its fabrication and use in a weapon. Neutrons from spontaneous fission of ^{240}Pu can cause pre-initiation in a weapon, making the yield uncertain. Gamma radiation from ^{241}Am (a product of the decay of ^{241}Pu), along with neutrons from ^{240}Pu, give rise to human exposure, but can be circumvented by shielding or using more workers.

Probably, the most important barrier is the radiation barrier provided by fission products. If this is absent in the immobilised plutonium, other barriers will have to be considerably augmented. The "self protecting" standard for spent fuel is usually taken to be 100 rad per hour at one meter. Depending on the physical security measures, this may not be sufficient for either spent fuel or immobilised plutonium. It takes an exposure of about 450 rads for a lethal dose (LD50, a 50 per cent chance of death). In the absence of guards or other measures one can imagine a scenario where spent fuel or immobilised plutonium is not self protecting. The dose rate one meter from the ends of a spent fuel assembly are reduced by a factor of 5 to 10 from those one meter from the midpoint. The dose rate at five meters from the midpoint is reduced by over an order of magnitude from that at one meter (60). Dose rates should be more than 1 000 rad per hour to be referred to as "self protecting". Fission products, such as ^{137}Cs, which provide the barrier have a half-life of about 30 years, after which other barriers will have to be augmented. It is possible that the plutonium could be sealed in an geologic repository in that time, replacing the radiation barrier by lack of accessibility, *i.e.* an isolation barrier.

Chemical dilution itself is not a practical barrier because chemical separation methods for plutonium are well known (61) and practiced in many countries. However, coupled with the radiation barrier, chemical separation becomes more difficult (dangerous), costly and visible. A low percentage of plutonium in the immobilisation material will likely be required because of performance and criticality considerations in a repository or borehole. This means that more packages would have to be stolen or diverted to obtain a given amount of plutonium.

Physical size and weight can be a significant impediment to the theft or diversion of packages containing immobilised plutonium, especially when coupled with a radiation barrier and physical protection measures. The minimum size and weight should be set so that heavy equipment such as cranes and fork lifts are necessary to move the packages. Glass logs being considered, for example, are about 0.6 meters in diameter and three meters long, and have a weight of about two tonnes.

The security of plutonium in a geologic repository or borehole depends on effective, and often synergistic, performance of the barriers of the disposal system. If immobilised plutonium is not provided with a radiation barrier, other barriers will have to be significantly enhanced. Currently, theft, rather than covert diversion by the host country, is the primary threat. As well as considering the on-site guards, as a response to this threat, the design and evaluation of a security system should include consideration of available backup forces, such as off-duty guards, local and regional police, and nearby military forces. If these "extra" forces are not available, then additional guards and physical protection measures and devices will be required on the site to reduce the risk of theft and provide a sufficient level of security.

Chapter 4

COUNTRY PROGRAMMES

4.1   Belgium

Rationale for decisions in the nuclear fuel cycle

The Belgian energy policy was focused for about 20 years (1974-1993) on the development of a significant nuclear power contribution to the electric generating capacity. About 5.75 GWe of nuclear capacity are currently on line, producing approximately 60 per cent of the total electricity output. Belgium did not build its own reprocessing capacity, although both the technology and the infrastructure (EUROCHEMIC) were available, but signed standing reprocessing contracts with COGEMA for a fraction of its spent fuel arisings.

From 1978 until 2000 a total amount of 670 t HM of spent fuel has been or will be reprocessed at COGEMA's La Hague plant. A total of approximately 4.8 t of plutonium will be returned to Belgium.

In December 1993 ended a parliamentary debate on reprocessing and on the use of MOX fuel in two of Belgium's nuclear power plants, Doel-3 and Tihange-2. The Belgian Parliament endorsed the use of MOX fuel and approved continued performance of the current reprocessing contracts signed in the late 1970s with France. Following the decisions taken by the government on the recommendation of parliament, the once-through option is to be placed on the same level as reprocessing. The next debate in parliament will take place in 1998 and should allow the government to decide on the future strategic options for the back-end of the fuel cycle. In the meantime, the reprocessing contracts signed after the late 1970s, for reprocessing during the period after 2000, are kept on hold during a 5-year period.

The current policy concerning the back-end of the fuel cycle will be pursued until 1998, when the situation will be re-assessed on economic, radiological and resource management grounds. The parliament will then make recommendations to the government on the future strategic options for the back-end of the nuclear fuel cycle. As long as the reprocessing option is held open, Belgium will consider plutonium, together with uranium, as a valuable energy resource.

Experience in nuclear and plutonium technology

Since the early days of nuclear energy (1960s), Belgium embarked on the development of fuel cycle technology and facilities. The nuclear research centre at Mol constituted the focal point around which a considerable nuclear industry has been set up.

- Belgonucléaire built in the 1970s a plutonium fuel fabrication plant for fast reactor fuel which was transformed in the 1980s into a LWR-MOX fuel fabrication plant. The

131

industrial production of MOX fuel which started in 1986 reached, since several years, its nominal capacity of 35 t HM per year. A doubling of the capacity is envisaged.

- FBFC has a PWR-UO$_2$ fuel fabrication capacity of 400 t per year (operational since the 60s) and has installed a MOX fuel assembly plant for the 35 tonnes of fuel pins produced at Belgonucléaire.

- Transnubel (a subsidiary of Belgonucléaire) is specialised in transportation of nuclear materials (plutonium, spent fuel and waste materials).

- Responsibility with waste treatment and disposal lies with the National Institute for Radioactive Waste and Fissile Materials (NIRAS/ONDRAF). It is a state owned organisation which owns the waste treatment plant BELGOPROCESS.

- BELGOPROCESS (former EUROCHEMIC) is the processing plant for all waste produced by the nuclear industry. It has limited facilities to treat alpha-contaminated waste. Important facilities have been constructed and operated for treatment and storage LLW, MLW and HLW (PAMELA).

- Waste disposal has been studied and developed at the Nuclear Research Centre from 1950 and is currently being pursued very actively. An underground laboratory (250 m deep) has been dug in compact clay. The Belgian Parliament has proposed to further develop the disposal option in the boom clay layer at Mol. Plutonium containing waste material can be adequately disposed of. A safety study (SAFIR) endorsed the clay option for HLW and alpha waste. Further studies are to be made in order to assess whether unreprocessed fuel could be disposed of.

4.2   Canada

Spent fuel produced by CANDU reactors operating in Canada contains a considerable quantity of plutonium because of the well known neutron economy and high conversion ratio of the reactor. Although much plutonium is produced per unit of electricity generated, because of the relatively low burn-up (about 8 MWd per kg U), of the natural uranium once-through cycle currently in use, the concentration in the spent fuel is low (about 0.27 wt. per cent fissile). Thus, given the present price of reprocessing, extraction of that plutonium to recycle through the reactor is not economically attractive.

Current practice is then to store the spent fuel in water filled storage bays at the power reactor sites. Canadian utilities' current on-site storage capacity for spent fuel allows time to develop an integrated disposal strategy and permits decisions on the ultimate fate of the fuel and of the plutonium it contains, to be deferred. Further deferral can be achieved by means of dry storage in relatively inexpensive concrete canisters after a suitable cooling time in the storage pools. The use of these concrete canisters is currently being demonstrated, and it is anticipated that the fuel can be stored in them for periods greater than 50 years.

A Nuclear Waste Management R&D programme led by the AECL has been active for more than 10 years with broad Canadian participation. The overall objective of this programme is to develop the technology to ensure that there will be no significant effects on the environment from these radioactive fuel wastes. An important activity of this programme has been to develop methods for safe, final disposal of spent fuel, either as the existing bundles or as vitrified waste from reprocessing, in the underground repositories cut into igneous rock formations or plutons. The proposed method and the methods used to demonstrate its efficacy are currently being considered by the Canadian Government in an extensive review which includes public participation.

In summary, it is presently being demonstrated that safe methods exist to store and eventually dispose spent fuel produced in Canada without the use of reprocessing. At the present time it appears that the most economic route would be to use this method of disposal. However, the decision to dispose the fuel in this way has not yet been taken, nor need it to be in the near future.

4.3   France

Status of the nuclear sector in France

The French electronuclear programme was launched at the time of the first oil shock with a double objective: to increase the French energy independence and to provide a competitive energy source. At the same time, an extensive energy conservation programme was set in motion.

Today, France has a standardized set of 55 Pressurised-Water Reactors (PWRs) with a power of nearly 60 000 MWe, which supplies 75 per cent of the country's electricity. These reactors operate with an availability factor of more than 80 per cent. Three other reactors are under construction.

France has strategically opted for reprocessing-recycling

Spent fuel unloaded from PWRs contains 1 per cent plutonium, 95 per cent uranium and 4 per cent fission products and minor actinides.

The French strategy has two principal objectives:

First, there is the requirement for good long-term management of natural resources which has long been a keynote of French policy. This option leads to savings of about 20 to 25 per cent of the initial natural uranium through a single recycling of the uranium and plutonium; this percentage can be increased to 90 per cent or more, through multiple recycling.

Second, it is generally accepted that the quantity of radioactive products that must be stored for thousands of years should be reduced as much as possible, in order to avoid leaving to future generations enormous quantities of plutonium and long-lived radioactive waste. After 10 000 years, plutonium will still represent more than 95 per cent of the radiotoxicity contained in the spent fuel. In order to limit the toxicity of the non-recycled products, the French system, which combines reprocessing as such with plutonium recycling, attains very high levels of plutonium recovery which substantially exceed 99.5 per cent.

Reprocessing and recycling of fissionable materials contribute to these objectives: the plutonium separated after reprocessing of the spent fuel can be recycled, either in the form of MOX in pressurised water reactors, or in fast neutron reactors used as breeders or incinerators. In addition, reprocessing operations allow optimum final packaging of the only real waste products (minor actinides and fission products) which no longer have energy potential.

The French approach to long-lived waste is defined by the law of 30 December 1991. This law provides for the diversification of research on high-level, long-lived waste management with three goals in mind:

- the separation and transmutation of long-lived waste;

- determining the possibilities of storage in deep geological formations, with the construction of underground research laboratories; and

- surface storage of waste and the resolution of resulting containment problems.

The law requires that, within 15 years, an evaluation report of this research be presented to the French Parliament, accompanied by a bill authorising the creation of a storage centre, should the need be established.

Reprocessing/recycling in France

Every year, about 1 100 t of spent fuel are unloaded from the reactors of the French electronuclear sector.

The UP2-COGEMA plant located at the Hague is capable of reprocessing 800 t of spent fuel each year, that is to say, to separate about eight tonnes of plutonium each year. This plant is presently used essentially for French needs, whereas the "twin" plant UP3 is used for the reprocessing of foreign fuel.

To avoid stockpiling of separated plutonium for which there is no immediate need, EDF has its spent fuel reprocessed only as and when openings arise for the extracted plutonium. In the short run, therefore, the quantity of fuel reprocessed depends not just on the capacity of the reprocessing plants, but on the capacity of the MOX manufacturing plants and the authorisations for the use of MOX fuel issued to French reactors.

At present (with the start-up of MELOX in 1995 and the new licences), it appears that MOX recycling will fully match reprocessing plant capacity in the future.

Recycling in the form of MOX in PWRs

Currently, nine French pressurised-water reactors are loaded with MOX fuel. Sixteen French EDF 900 MWe PWRs were licensed, from the very beginning, to recycle plutonium. Recently, EDF has initiated administrative operations to license twelve additional 900 MWe PWRs to recycle plutonium. The MOX fuel used comes either from the COGEMA plant at Cadarache, with a capacity of 30 t per year, or from the Belgonucléaire plant at Dessel (Belgium) with a capacity of 35 t per year.

The MELOX plant, built by COGEMA at Marcoule, started up in 1995. Its capacity will be between 120 and 160 tonnes of fuel and may be subsequently increased. From the eight or so tonnes of plutonium that UP2 can separate each year, MELOX will produce 20 or more reloads for the needs of EDF, which foresees 28 MOX licensed reactors by the end of the decade.

A reactor, 30 per cent of whose fuel is in the form of MOX, both produces and consumes plutonium; the annual operational balance of a MOX loaded reactor is a net production of about 30 to

60 kg of plutonium per year, or about 5 to 10 kg of plutonium per TWh, whereas a conventional fuel reactor produces 180 to 220 kg of plutonium per year, or about 30 to 40 kg of plutonium per TWh.

Table 18 **Net production (production less consumption) of total plutonium in a PWR**

Total Plutonium	Present PWRs	Present PWRs	Present PWRs with 30% MOX	Present PWRs with 30% MOX
Burn-up (MWd/t)	33 000	50 000	33 000	50 000
kg Pu/year	+ 220	+180	+ 60	+ 25
kg Pu/TWh	+ 40	+ 30	+ 10	+ 5

Depending on the economic conditions, France would orient future industrial techniques in the direction of greater or lesser consumption of plutonium, the basic objective being, in all cases, rigorously maintaining the separated stocks at the necessary industrial level.

With this in mind, research is being conducted to further decrease the net plutonium production of a MOX fuel loaded reactor, or even to make it negative, should it be shown to be necessary, either by increasing the proportion of MOX elements in the reactors, or by increasing the proportion of plutonium in these elements.

Recycling in the form of MOX in FRs

Initially, fast neutron reactors were designed to operate in the breeder mode, that is to say, to produce more plutonium than they consume.

The present relaxed situation in the uranium market and the need to regulate the quantities of plutonium produced justify a high priority study of the possibility of burning any type of plutonium in fast neutron reactors. This is the CAPRA programme conducted by the CEA, with which many countries, including Japan, are associated. The objective aimed at is a net consumption of plutonium of between 80 and 110 kg plutonium per TWh.

Superphénix will contribute to this programme: the progressive conversion of its core from the breeder mode to the sub-breeder mode will allow the demonstration of the flexibility of the design and will provide the first industrial experience with a fast neutron reactor as a net consumer of plutonium. Starting with a core which produces about 25 kg plutonium per TWh, the sub-breeder mode will make it possible to arrive at a net consumption of at least 20 kg per TWh of plutonium after the end of the decade.

Table 19 **Net production (production less consumption) of total plutonium in Superphénix**

Total Plutonium	1st core 1994-1995	1st modified core 1995-1998	2nd core 1999-2002	3rd core 2003 on
kg Pu/year	+ 180	+ 90	+ 0	- 150
kg Pu/TWh	+ 25	+ 12	+ 0	- 20

Ultimately, depending on the results obtained with the CAPRA programme and with Superphénix, France can look forward to the construction of a mixed system composed of pressurised-water reactors, operating either with conventional fuel or MOX, and fast neutron reactors. Thus, it will be possible to manage the total quantity of plutonium, whether separated or contained in the irradiated fuel.

Conclusions

Today the use of plutonium in French electronuclear plants is an industrial reality which responds to national concerns with respect to the rational management of raw materials, the exploitation of the energy potential of these materials and the limitation of environmental impact. The surveillance of an international control system (such as EURATOM or the IAEA), should preclude nuclear proliferation problems.

4.4   Germany

Utilisation of plutonium by recycling in LWRs is practised in Germany since 1968. The amendment of the German Atomic Law in 1994 has also opened the long-term intermediate storage with subsequent direct disposal of spent fuel as an equal option in addition to the reprocessing/recycling route. Both alternatives, the closed and open fuel cycle routes, will be exercised by German utilities and the most economical option will be adopted under long-term considerations.

Since 1968, MOX fuel assemblies have been fabricated at the SIEMENS MOX fuel fabrication plant at Hanau (formerly Alkem) and the throughput has been increased up to 25 t HM per year. More than 150 t HM of MOX fuel for BWRs and PWRs, and 5 t HM for fast breeder reactors have been produced and delivered. In 1992, the old MOX plant was closed. Construction of the new MOX fuel fabrication plant in Hanau, with a capacity of 120 t HM per year, started in 1987 and the operation licence was granted in March 1991. The plant was completed up to 95 per cent in 1992, but its final completion and start-up operation was blocked politically by the State Government of Hesse. Although the Federal Government, SIEMENS, and the German utilities continued to support this project, the final completion and the start-up of the plant could not be realised due to the stringent anti-nuclear policy of the State Government of Hesse. Therefore, SIEMENS and the German utilities were forced politically in mid-1995 to abandon the plant. For the future, the German utilities will have their plutonium fabricated into MOX assemblies at the plants of COGEMA, BNFL and Belgonucléaire, under fuel supply contracts concluded with SIEMENS and Fragema. The plant operators make every effort to reduce the plutonium surplus stocks and the bottleneck for the supply of MOX assemblies resulting from the abandonment of the Hanau MOX plant.

In Germany, recycling of plutonium in BWRs and PWRs has proved quite satisfactory since 1968. Currently, twelve nuclear power plants are licensed for the use of MOX assemblies, with a total capacity of 100 t HM per year. An application for MOX use has been filed for another five nuclear power plants.

Reprocessed uranium arising under reprocessing contracts is converted into uranium oxide and put into intermediate storage, as its use is presently not considered economical. This practise would be reviewed as economic conditions for uranium and enrichment services evolve in the future. The use of fuel assemblies containing reprocessed uranium has already been demonstrated in Germany,

136

with one assembly loaded in the Obrigheim reactor and eight assemblies loaded in the GKN reactor in 1986 and 1987.

4.5   Japan

In Japan, a country not blessed with energy resources, it is indispensable to plan for its energy security based on the future outlook in order to maintain and further develop its economic and social activities. Japan intends to guarantee its future energy security by steadily carrying forward research and development aimed at the commercialisation of nuclear fuel recycling involving the reprocessing of spent fuel and the recovery of plutonium, uranium, etc. in order to re-use them as nuclear fuel. Furthermore, recycling of nuclear fuel is meaningful in terms of sparing resources, preserving the environment and also contributing to improved management of radioactive waste.

To put it concretely, Japan maintains the basic idea of utilising Fast Breeder Reactors (FBRs) for the mainstream of nuclear power generation after a long period, during which the Light-Water Reactors (LWRs) are used together with FBRs, and intends to steadily carry forward research and development in stages, under co-operation between the government and the private sector, for the purpose of establishing a technological system of nuclear fuel recycling based on FBRs aimed at commercialisation by about 2030. On the other hand, from the standpoint of establishing a comprehensive technological system concerning the use of plutonium that will be needed in the timeframe of the FBRs in the future, and long-term overall improvement of the economic efficiency of nuclear fuel recycling, is important to carry out nuclear fuel recycling on a certain scale. Therefore, Japan intends to build commercial reprocessing plants in order to gain experience with their operation and to implement nuclear fuel recycling based on existing LWRs.

Regarding the economic efficiency of the nuclear fuel cycle, although at the present time it is estimated that the use of MOX fuel in LWRs will be somewhat more costly than directly disposing of spent fuel, there is effectively no difference from the point of view of total power generation costs. Japan intends to work on improving the fuel cycle economic efficiency from a long-term perspective, by means of standardization of fuel specifications, etc. Regarding the FBR nuclear fuel cycle, Japan is expecting that it will achieve such a high efficiency as with LWRs, through the introduction of innovative technology and other efforts.

Furthermore, Japan's nuclear fuel recycling programme is implemented on the basis of the principle of not storing more plutonium beyond the amount required to support the programme, *i.e.* the principle of having no surplus plutonium, as well as having very strict management of nuclear materials. In addition it will work on the basis of rational and consistent plans and strive for transparency in order not to give rise to any international doubts concerning violation of non-proliferation agreements of nuclear power.

4.6   Netherlands

In the Netherlands, spent fuel arising from the two nuclear power plants is reprocessed in France and the United Kingdom. The owners of the Dodewaard nuclear power plant and the Borssele nuclear power plant have concluded reprocessing contracts with BNFL (Sellafield) and with COGEMA (La Hague).

137

In the past a small demonstration loading of MOX assemblies took place at the Dodewaard nuclear power plant. Further use of MOX fuel, however, is not foreseen.

4.7   Russia

Russia began using plutonium as a nuclear fuel in the second half on the 1950s. In 1957 a core of metallic plutonium alloy was fabricated for the pulsing fast reactor IBR-30. In 1959-65 plutonium dioxide fuel was made for the fast experimental reactor BR-5.

Systematic studies of plutonium fuel began in 1970 in the BOR-60 reactor. That work eventually conformed to the fast breeder reactor using plutonium as the most efficient way to expand nuclear power fuel resources.

Two cores containing weapons-grade plutonium have been tested in the BR-10 experimental fast reactor. Large batches of MOX fuel pins made of different technologies with plutonium of various isotopic compositions have been tested in the BOR-60 research fast reactor. This reactor has been operated for many years recycling plutonium.

With continued successful MOX fuel tests, the scope of the tests were expanded to the prototype sodium-cooled fast reactors BN-350 and BN-600 which have been fueled from the very beginning with enriched uranium. The semi-industrial facilities "Granat" and "Paket", capable of producing up to ten plutonium fuel sub-assemblies annually for these reactors, were built at the PO Mayak at Chelyabinsk to experimentally substantiate the technology of plutonium utilisation in fast reactors. In the BN-350 reactor the tests have been performed with the subsequent investigation and chemical reprocessing of test fuel sub-assemblies with MOX fuel (350 kg of weapons-grade plutonium). More than 2 000 such fuel elements have been assembled and tested.

Post-reactor investigation showed that fuel element endurance was not exhausted with a burn-up of 9-11 per cent h.a. MOX fuel fabrication technologies that produced less dust than the mechanical mixing of oxides were studied and developed to improve radiation characteristics. The sol-gel fabrication process of granulated MOX fuel was used for the subsequent pellet pressing. However, this process has not provided very high or stable pellet quality. Therefore, a method of ammonia co-precipitation of uranium and plutonium was developed in parallel. Most of these twelve fuel sub-assemblies manufactured for the BN-600 reactor passed in-pile tests.

Vibrocompacted fuel has been studied using granulated fuel obtained through various means. Among them is a method of electrochemical uranium oxide and plutonium oxide co-precipitation. All fuel elements loaded into the BOR-60 reactor have been built using this technology. Also, sub-assemblies have been tested in the BN-350 and the BN-600 reactors.

MOX fuel development for the BN-type fast breeder reactor has progressed to the initial design and construction of two units of BN-800 power fast reactors at the South Urals and Beloyarskaya sites. The construction of a MOX fuel plant, called Complex-300, is 50 per cent complete. It could produce up to 900 sub-assemblies per year as fuel for these reactors. The reactors are in the initial stage of construction. Currently, both projects are suspended. The pilot installations at the Mayak "Granat" and "Paket" will be back-fitted to meet the latest safety and environmental requirements of the state nuclear surveillance commission, Gosatomnadzor.

The Russian closed nuclear fuel cycle was first achieved with the commissioning of the RT-1 reprocessing plant in 1976 at Mayak near Chelyabinsk. The plant capacity is 400 t per year of the main fuel type, that for the VVER-440 reactor. To date about 3 000 t of spent fuel have been reprocessed in this plant.

The main product of the RT-1 spent fuel reprocessing is enriched uranium. However, the plant can also extract as much as 2.6 t of civil plutonium per year. Last year's plutonium production dropped to 0.6 t. The total quantity of extracted civil plutonium stored at the Mayak site is about 30 t. This plutonium is stored reliably as dioxide to be used in fast reactors.

A new radiochemical plant, RT-2, is under construction not far from Krasnoyarsk. Its main objective will be to reprocess VVER-1000 spent fuel. Its first phase, to be put in operation after 2005, will reprocess 1 500 t of fuel per year.

The accumulation of the 30 t of plutonium at Mayak, the plutonium to be extracted from spent fuel at the RT-2 plant, along with the 100 t of ex-weapons plutonium expected to be released, aggravate the problem of plutonium management. There is no doubt that, in the near term, management measures must be taken for reliable plutonium storage. In addition to the storage of civil plutonium, storage for released ex-weapons plutonium should also be arranged. At the same time, such storage should not be utilised for very long time due to economic, environmental and political reasons.

The plutonium management options, with emphasis on the extracted civil, and the expected to be released, ex-weapons plutonium, are being analysed in the Ministry for Atomic Energy of Russia. They are based on the following principles:

- Russian experience of plutonium management should be maximised;

- plutonium diversion resistance should be considered an important factor;

- plutonium management should be economically and environmentally acceptable; and

- separated plutonium management options should serve as a good base for the development of the optimal fuel cycle under long-term perspectives.

The concept of Nuclear Power Centres (NPC) comprising reprocessing plants, MOX fuel production plants and plutonium-fueled reactors could meet the above mentioned principles. The first of this kind of NPC is supposed to be the "Mayak" center at Chelyabinsk which includes the RT-1 reprocessing plant, currenlty in operation, and where the construction of the MOX fabrication facility COMPLEX-300 and the three fast reactors BN-800 has been planed.

Once-through MOX fuel burning in BN-800 reactors is considered as the first stage of plutonium management. It is aimed to rapidly convert separated plutonium into spent fuel in order to minimise the risk of plutonium proliferation. The next step envisages the recycling and the burning of the surplus plutonium originating from spent fuel including MOX fuel. New fast reactor core designs are considered without breeding zones, having increased plutonium content in MOX fuel. New fuel composition on the base of an inert matrix replacing ^{238}U is under consideration.

As regards plutonium utilisation in thermal reactors, Russia is presently also analysing this option. Studies underway address the advantages and disadvantages of plutonium use in VVER-1000 reactors as compared to fast reactors. A critical assembly, called SUPR, is being constructed at IPPE in Obninsk. It will be used to study the safety characteristics of VVERs using plutonium, including weapons-grade material.

4.8   Switzerland

Currently, five nuclear power plants with a combined power output of 3 GWe are in operation in Switzerland, yielding, on average, a total of about 0.5 t of fissile plutonium in spent fuel per year. Spent fuel is partly sent to France and the UK for reprocessing and is partly stored on site for later disposition.

A total of about 3 000 t of spent fuel is expected to be unloaded during the 40-year planned life of these five reactors. About one third of the spent fuel is currently contracted for reprocessing. No decision has yet been taken on the remaining two thirds of spent fuel arisings, but it is expected that the existing commitments to reprocessing will be honoured in full.

The total amount of spent fuel committed to reprocessing will yield 6.0 to 6.5 t of fissile plutonium. About 1.5 t of this material has to-date been recycled as MOX fuel in the Beznau-1 and Beznau-2 reactors. The Gösgen nuclear power plant is planning to start recycling plutonium in 1997 and the Leibstadt nuclear power plant will follow in about 2001.

4.9   United Kingdom

Status of the industry

Reprocessing of irradiated fuel in the UK has been carried out since 1952 at BNFL's Sellafield site. The Magnox reprocessing plant has been operating since 1964. Since the start of operations, over 35 000 t U of Magnox fuel has been reprocessed and the plant, with some refurbishment, is expected to continue operation through to the end of the Magnox programme sometime within the next 10 to 15 years.

The THORP plant for the reprocessing of AGR and LWR oxide fuels commenced operation in 1994 and is programmed to reprocess some 7 000 t U in the first 10 years of operation. Plutonium is recovered from both of these reprocessing operations in the form of PuO_2.

Fuel from the Prototype Fast Reactor (PFR) at Dounreay has been reprocessed in a mixed oxide reprocessing plant at Dounreay since 1979, with the plutonium arisings transferred to Sellafield for storage. On completion of PFR reprocessing in some 6 to 7 years time a total of around 50 t of fuel will have been reprocessed.

Currently, there are some 40 t of separated safeguarded UK plutonium in store at Sellafield and a further 25 t or so, are expected to arise by the end of the Magnox programme. Plutonium arising from the contracted reprocessing of AGR fuel in THORP together with those arisings from overseas LWR fuel to which BNFL has title and the arisings from the UKAEA's WAGR and SGHWR fuel will be some 15 t to 20 t.

Plutonium at Sellafield and Dounreay (whether separated or in spent fuel) is stored safely and securely and under international safeguards.

Government policy

The International Atomic Energy Agency (IAEA) circular INFCIRC/225 (Rev. 2) sets out internationally agreed guidelines for physical protection arrangements and the Convention on the Physical Protection of Nuclear Materials deals specifically with the international transport of plutonium. The guidelines established by INFCIRC/225 (Rev. 2) and the provisions of the convention are accepted by the UK and the governments of all its reprocessing customers. Under these, plutonium will be subject to adequate physical protection measures in the UK, during transport to reprocessing customers and when in the possession of reprocessing customers. As a condition of exporting plutonium the UK Government requires an assurance from the recipient state that it will apply the appropriate physical protection measures. It also requires assurances as to the non-explosive use of the plutonium, the application of safeguards and the non-retransfer of the material to third parties. Assurance on these measures, and on physical protection are set out in relevant Nuclear Co-operation Agreements and exchanges of notes.

The UK Government has stated that within the relevant national and international framework (including safeguards, regulatory matters and applicable national policies), it is for the owners of plutonium to make commercial decisions on recycling.

Plutonium utilisation

Historically, UK policy on plutonium utilisation was based on an assumption of early commercial exploitation of fast reactors. In 1988 the UK Government gave notice of withdrawal of financial support for operation of the Prototype Fast Reactor at Dounreay after 1994 and terminated its support for the European Fast Reactor Project in March 1993. This has effectively postponed any deployment of fast reactors commercially in the UK for the foreseeable future.

The prospects for recycle of plutonium in thermal reactors remains under review.

The MDF (MOX Demonstration Facility) situated at Sellafield started operation in 1993. The plant has a 8 t HM per annum LWR MOX capacity for which the majority has already been contracted.

BNFL's Sellafield MOX Plant (SMP) is under construction and is due to operate in 1997 with a capacity of 120 t HM per annum for LWR MOX fuel. This plant is situated adjacent to the THORP reprocessing facility and is designed to utilise plutonium from THORP that may have been stored for several years; it will also be capable of receiving and using plutonium from other sources.

4.10   United States

The Administration's non-proliferation policy states that the United States does not encourage the civil use of plutonium and, accordingly, does not itself engage in plutonium reprocessing for either nuclear power or nuclear explosive purposes. The United States, however, will maintain its existing commitments regarding the use of plutonium in civil nuclear programmes in Western Europe and Japan. In addition, the policy commits the United States to explore means to limit the stockpiling of plutonium from civil nuclear programmes and to seek to minimise the civil use of highly-enriched uranium.

Accordingly, low-enriched uranium will continue to be used in commercial reactors in a once-through fuel cycle with spent fuel planned for geological disposal in an approved repository.

Chapter 5

CONCLUSIONS

Quantities of separated plutonium, which have a variety of isotopic compositions, have been increasing over the past three decades and are expected to continue to increase during the next few years. It is estimated that civil plutonium quantities will exceed 180 tonnes, on a cumulative basis, by the end of the century. According to the International Atomic Energy Agency, the evolution of plutonium inventories over the coming decades is subject to significant uncertainties. Successful implementation of mixed uranium-plutonium oxide (MOX) recycling programmes, which are under way in a number of countries, would result in an equilibrium between plutonium separation and consumption in OECD countries and would eventually reduce civil plutonium stocks.

This report dealt with and was restricted to the technical options for the management of separated plutonium. A number of complex and interrelated non-technical factors, which have not been addressed by the expert group, such as considerations of national and international policy concerns: non-proliferation, public acceptability, economics, environmental impact and infrastructure would inevitably play a central role in thoroughly developing, implementing and completing the technical options examined in this study.

MOX fuel has been manufactured on a semi-industrial scale for nearly 25 years and has been recycled in several thermal and fast reactors. The existing MOX fuel fabrication facilities in OECD countries for thermal reactors can produce some 190 tonnes HM per year which requires some 10 to 12 tonnes plutonium. This quantity is required for the present number of 32 reactors licensed to use MOX fuel. MOX fabrication capacity is expected to double around the year 2000. Production and recycling of MOX in Light-Water Reactors (LWRs) has reached industrial maturity during the last ten years, as is evidenced by the recent large LWR recycling programmes. Today, considerable experience has already been accumulated.

In the next 15 to 20 years:

- In a number of OECD countries, a significant part of the separated plutonium will be recycled as MOX in thermal reactors. The temporary plutonium accumulation will need to continue to be stored and some will need to be purified in order to have the internally arising americium removed before recycling. All parts of the fuel cycle are safeguarded, including stored plutonium. The addition and expansion of existing industrial capacities will enable the reduction of stocks to the level required for the efficient operation of the facilities.

- Existing technologies for storage and recycling developed in the civil fuel cycle can cope, if necessary, with surplus material produced from non-civil sources.

In the longer term, depending on the evolution of nuclear policies, the following alternatives (alone or in combination) may be considered:

- a continuation of the current management techniques, as quoted above;

- plutonium burning in fast neutron reactors (which could, if desired, be used also for breeding) and other dedicated reactors, including thermal reactors; and

- conditioning plutonium to a form compatible with regulatory requirements for final disposal.

As shown in this report, the required technologies for the short and medium terms (15 to 20 years) are both commercially proven and available in OECD countries. Such technologies are safe and properly safeguarded.

The following alternative technologies for the longer term would still have to be fully demonstrated through continuous research and development efforts:

- burning plutonium in fast and other dedicated reactor types; and

- conditioning plutonium to a form compatible for final disposal.

The management of separated plutonium presents no major technical difficulties, but is merely a matter of applying existing technology to the minimisation of any plutonium stocks.

REFERENCES

1. OECD Nuclear Engery Agency, *Plutonium Fuel: an Assessment*, 1989, Paris.

2. OECD Nuclear Energy Agency, *The Safety of the Nuclear Fuel Cycle*, 1993, Paris.

3. OECD Nuclear Energy Agency, *The Economics of the Nuclear Fuel Cycle*, 1994, Paris.

4. OECD Nuclear Energy Agency, *Physics of Plutonium Recycling,* Paris.

    Vol. I: Issues and Perspectives, 1995.
    Vol. II: Plutonium Recycling in Pressurized-Water Reactors, 1995.
    Vol. III: Void Reactivity Effect in Pressurized-Water Reactors, 1995.
    Vol. IV: Fast Plutonium-Burner Reactors, 1996.
    Vol. V: Plutonium Recycling in Fast Reactors, 1996.
    Vol. VI: Multiple Recycling in Advanced Pressurized-Water Reactors, in preparation.

5. IAEA, *Problems Concerning the Accumulation of Separated Plutonium*, report of an Advisory Group meeting held in Vienna, 26-29 April 1993, IAEA-TECDOC-765, September 1994, Vienna.

6. Semenov, B., Oi, N., Grigoriev, A., Takats, F., "Spent Fuel Management Overview", *International Symposium on Spent Fuel Storage – Safety, Engineering and Environmental Aspects*, IAEA/NEA, 10-14 October 1994, Vienna.

7. IAEA, *Year Book 1996*, September 1996, Vienna.

8. IAEA, *Strategies for the Back-End of the Nuclear Fuel Cycle*, proceedings of a Technical Committee meeting held in Vienna, 1-4 June 1993, IAEA-TECDOC-839, November 1995, Vienna.

9. US National Academy of Sciences, *Management and Disposition of Excess Weapons Plutonium*, 1994, Washington, DC.

10. *Statement by the United States of America to the 39 regular session of the General Conference of the IAEA*, 18-22 September 1995, Vienna.

11. US Department of Energy, *Technical Summary Report for Surplus Weapons-Usable Plutonium Disposition,* July 1996, Washington, DC.

12. IAEA, *Safe Handling, Transport and Storage of Plutonium*, proceedings of a Technical Committee meeting held in Vienna, 18-21 October 1993, IAEA-TECDOC-766, October 1994, Vienna.

13. "Plutonium in the Community", article written by members of the European Commission DG XVII/C-3, *Energy in Europe*, No. 22, December 1993.

14.   Nuclear Energy Institute, *Proceedings of the Fuel Cycle 94 Conference*, 20-23 March 1994, Boston.

15.   Parkes, P., "Packaging and Storage of Plutonium in BNFL's THORP Complex", *Storage in the Nuclear Fuel Cycle,* Institute of Mechanical Engineers, 18-19 September 1996, Manchester.

16.   US Department of Energy, *Criteria for Safe Storage of Plutonium Metals and Oxides*, US DOE standard 3013, December 1994, Washington, DC.

17.   US Department of Energy, *Assessment of Plutonium Safety Storage Issues at DOE Facilities*, US DOE/DP 0123T, January 1994, Washington, DC.

18.   US Department of Energy, *Plutonium Working Group Report on Environmental, Safety and Health Vulnerabilities Associated with the Department's Plutonium Storage*, US DOE EH-415, September 1994, Washington, DC.

19.   US Department of Energy, *Environmental Analysis of Sea Shipments of Plutonium from Europe to Japan*, ANL/IEP-88-50.

20.   Bairiot, H., Deramaix, P., "MOX Fuel Development: Yesterday, Today and Tomorrow", *Journal of Nuclear Materials*, No. 188, 1992, pp. 10-18.

21.   Bairiot, H., "Laying the Foundations for Plutonium Recycle in Light-Water Reactors", *Nuclear Engineering International*, Vol. 29, No. 350, January 1984.

22.   US NRC, *Final Generic Environment Statement on the Use of Recycled Plutonium in Mixed Oxide Fuel in Light-Water Cooled Reactors*, NUREG-0002, August 1976, Washington, DC.

23.   Haas, D., Vandergheynst, A., van Vliet, J., Lorenzelli, R., Nigon, J.-L., "Mixed-Oxide Fuel Fabrication Technology and Experience at the Belgonucléaire and CFCa Plants and Further Developments for the MELOX Plant", *Nuclear Technology*, Vol. 106, No. 1, April 1994, pp. 60-82.

24.   Pay, A., Vandergheynst, A., "MOX Fuel Fabrication Plants in Dessel: Operating Experience with P0; P1, a Second Generation Plant", RECOD, April 1994, London.

25.   Trauwaert, E., Mostin, N., Lefèvre, R., "MOX Fabrication Experience at Dessel", *Proceedings of a Technical Committee meeting on Recycling of Plutonium and Uranium in Water Reactor Fuels*, IAEA, 1989, Cadarache.

26.   Bairiot, H., Deramaix, P., Vanderborck, Y., Mostin, N., Trauwaert, E., "Foundations for the Definition of MOX Fuel Quality Requirements", *Journal of Nuclear Materials,* Vol. 178, No. 2/3, February 1991, pp. 187-194.

27.   van Vliet, J., Pelckmans, E., Mostin, N., Lorenzelli, R., Moulard, M., Mares, C., "Industrial Qualification of TU2 Powder Use for MOX Fuel Fabrication", *Nuclear Europe Worldscan*, No. 11-12, 1994, p. 50.

28.   Bariteau, J.R., Heyraud, J., Nougues, B., "The MOX Fabrication at the CEA Cadarache Complex", *Proceedings of a Technical Committee meeting on Recycling of Plutonium and Uranium in Water Reactor Fuels*, IAEA, 1989, Cadarache.

29.   Charle, Th., Darbouret, B., "MOX Fuels, the Future MELOX Plant, Design and Safety Issues", *Symposium on the Safety of the Nuclear Fuel Cycle*, BNS-OECD/NEA, 3-4 June 1993, Brussels.

30.   Krellmann, J., "Plutonium Processing at the SIEMENS Hanau Fuel Fabrication Plant", *Nuclear Technology*, Vol. 102, No. 1, April 1993, pp. 18-28.

31.   Brähler, G., "Present Status of the Hanau MOX Fuel Fabrication Facility with Emphasis on Safety Issues", *Symposium on the Safety of the Nuclear Fuel Cycle*, BNS-OECD/NEA, 3-4 June 1993, Brussels.

32.   Schneider, V., Güldner, R., Brähler, G., "MOX Fuel Fabrication in the Federal Republic of Germany", *Proceedings of a Technical Committee meeting on Recycling of Plutonium and Uranium in Water Reactor Fuels*, IAEA, 1989, Cadarache.

33.   Edwards, J., Hexter, B.C., Powell, D.J., "Plutonium – Out of the Stockpile and into the MOX Market", *ATOM,* No. 427, March-April 1993.

34.   Macdonald, A.G., "The MOX Demonstration Facility – the Stepping-Stone to Commercial MOX Production", *Nuclear Energy*, Vol. 33, No. 3, June 1994, pp. 173-177.

35.   MacLeod, H.M., Yates, G., "Development of Mixed-Oxide Fuel Manufacture in the United Kingdom and the Influence of Fuel Characteristics on Irradiation Performance", *Nuclear Technology*, Vol. 102, No. 1, April 1993, pp. 3-17.

36.   Information received from Mr. T. Mishima, Nuclear Fuel Cycle Engineering Division, PNC, 16 June 1994.

37.   Fournier, W., Mouroux, J.P., "MELOX Progress Status", RECOD, April 1994, London.

38.   Young, M.P., "Commercial MOX Fuel Production – the Challenge", *Nuclear Energy*, Vol. 33, No. 3, June 1994, pp. 189-191.

39.   Schlosser, G.J., Manzel, R., "Recycling of Plutonium in Light-Water Reactors. Demonstration Programs", *Siemens Forschungs und Entwicklungsberichte,* Vol. 8, No. 2, 1979, pp. 108-113.

40.   van Vyve, J., Resteigne, L., "Introduction of MOX Fuel in Belgium NPPs – from Feasibility to Final Implementation", BNES Seminar: Fuel Management and Handling, *Nuclear Energy*, Vol. 35, No. 5, October 1995, pp. 307-310.

41.   Schlosser, G.J., Winnik, S., "Thermal Recycling of Plutonium and Uranium in the Federal Republic of Germany: Strategy and Current Status", *Back-End of the Fuel Cycle: Strategies and Options*, proceedings of a Symposium, IAEA/NEA, IAEA-SM-294/33, 11-15 May 1987, Vienna.

42.   Schlosser, G.J., Krebs, W.-D., Urban, P., "Experience in PWR and BWR Mixed Oxide Fuel Management", *Nuclear Technology,* Vol. 102, No. 1, April 1993, pp. 54-67.

43.   Rome, M., Salvatores, M., Mondot, J., Le Bars, M., "Plutonium Reload Experience in French Pressurized Water Reactors", *Nuclear Technology,* Vol. 94, No. 1, April 1991, pp. 87-98.

44. Barral, J.C., Hervouet, C., Lam-Hime, M., Bergeot, M.A., Larderet, P., Lefebvre, J.C., Vassalo, A., "French PWR Operation Feed-Back Comparison between Predicted and Measured Core Physics Parameters", *PHYSOR 90 International Conference on the Physics of Reactors: Operation, Design and Computation*, Vol. 1, 23-27 April 1990, Marseille.

45. Haas, D., "Building a Data Base on MOX Performance in LWRs", *Nuclear Engineering International*, Vol. 32, No. 391, February 1987, p. 35.

46. Goll, W., Fuchs, H.-P., Manzel, R., Schlemmer, F.U., "Irradiation Behavior of UO_2/PuO_2 Fuel in Light-Water Reactors", *Nuclear Technology*, Vol. 102, No. 1, April 1993, pp. 29-46.

47. Murogov, V., IPPE, private communication, 1995.

48. Bernard, C., Miquel, P., Viala, M., "Advanced Purex Process for the New Reprocessing Plants in France and in Japan", RECOD, 1991, Sendai.

49. Horner, D.E., Crouse, D.J., Kappelman, F.A., Goode, J.H., "Promotion of PuO_2 Dissolution in Nitric Acid with Cerium", *Transactions of the American Nuclear Society*, Vol. 21, 1975.

50. Harmon, H.D., "Dissolution of PuO_2 with Cerium (IV) and Fluoride Promoters", *DP--1371*, October 1975.

51. Bray, L.A., Ryan, J.L., in *Actinide Recovery from Waste and Low-Grade Sources*, Navratil, J.D., Schulz, W.W., eds., Harwood Academic Publishers, 1982, New York.

52. Koehly, G., Bourges, J., Madic, C., Lecomte, M., French Patent 2 561 314 EN 84 0 4764 1984, 1984.

53. Dastur A.R., Buss, D.B., "The Influence of Lattice Structure and Composition on the Coolant Void Reactivity in CANDU", *Proceedings of the 11th Annual Conference of the Canadian Nuclear Society*, June 1990, Toronto.

54. McKibben, J.M., et al., *Vitrification of Excess Plutonium*, WSRC-RP-93-755, Westinghouse Savannah River Company, May 1993.

55. Jostsons, A., Vance, E.R., Mercer, D.J., Oversby, V.M., "Synroc for Immobilizing Excess Weapons Plutonium", *Proceedings of XVIII International Symposium on the Scientific Basis for Nuclear Waste Management*, 23-27 October 1994, 1995, pp. 775-781, Kyoto.

56. Vance, E.R., Begg, B.D., Day, R.A., Ball, C.J., "Zirconolite-rich Ceramics for Actinide Wastes", *ibid*.

57. Laidler, J.J., "Conditioning of Spent Nuclear Fuel for Permanent Disposal", *Technology: Journal of the Franklin Institute*, Vol. 331A, 1994, pp. 173-181.

58. McPheeters, C.C., Pierce, R.D., Mulcahey, T.P., "Pyroprocessing Oxide Spent Nuclear Fuels for Efficient Disposal", *Proceedings of the ANS Topical Meeting on DOE Spent Nuclear Fuel*, American Nuclear Society, 1994, pp. 275-281.

59. US Department of Energy, *Management of Commercially-Generated Radioactive Waste, Environmental Impact Statement*, 1980, Washington, DC.

60. Lloyd, W.R., Sheaffer, M.K., Sutcliffe, W.G., *Dose Rate Estimates from Irradiated Light-Water Reactor Fuel Assemblies in Air*, UCRL-ID-115199, Lawrence Livermore National Laboratory, 31 January 1994.

61. Wick, O.J., ed., *Plutonium Handbook, a Guide to the Technology*, Vols. I and II, American Nuclear Society, 1980.

LIST OF EXPERT GROUP MEMBERS

AUSTRALIA

Dr. G. DURANCE   Australian High Commission, London

BELGIUM

Mr. L. BAEKELANDT	ONDRAF/NIRAS
Dr. L.H. BAETSLE	SCK/CEN
Mr. H. BAIRIOT	FEX
Mr. P. DERAMAIX	Belgonucléaire
Mr. P. VERBEEK	SYNATOM

CANADA

Dr. A. DASTUR   Atomic Energy of Canada Ltd.

FRANCE

Mr. J.-G. DEVEZEAUX DE LAVERGNE	COGEMA
Mr. M. MOULIE	Électricité de France
Prof. M. SALVATORES	Commissariat à l'énergie atomique/CEN-Cadarache
Mr. B. SICARD	Commissariat à l'énergie atomique/DCC

GERMANY

Dr. C.A. DUCKWITZ	SIEMENS AG
Mr. G. HOTTENROTT	RWE Energie

IRELAND

Mr. F.J. TURVEY   Radiological Protection Institute

ITALY

Dr. C. ARTIOLI	ENEA
Dr. G. DOMINICI	ENEA

JAPAN

Mr. H. MATSUO	Atomic Energy Bureau, STA
Mr. Y. MIKI	Power Reactor & Nuclear Fuel Development Corp.
Mr. T. MISHIMA	Power Reactor & Nuclear Fuel Development Corp.
Mr. H. MIZUMA	Atomic Energy Bureau, STA
Mr. K. SHIRAHASHI	Atomic Energy Bureau, STA
Mr. K. TAKAHASHI	Power Reactor & Nuclear Fuel Development Corp.

KOREA (REP. OF)
Dr. H.-K. JOO   Korean Atomic Energy Research Institute
Dr. Y.-J. KIM   Korean Atomic Energy Research Institute
Dr. B.-W. LEE   Korean Atomic Energy Research institute
Dr. D.-S. SOHN   Korean Atomic Energy Research Institute
Dr. D.-D. SUL   Ministry of Science and Technology

NETHERLANDS
Mr. C.J. JOSEPH (**Chairman**)   Ultra-Centrifuge Nederland NV

NORWAY
Dr. A. HANEVIK   Institute Energy Technology,
OECD Halden Reactor Project

RUSSIAN FEDERATION
Dr. A.N. CHEBESKOV   Institute of Physics and Power Engineering
Dr. E. KUDRIAVTSEV   MINATOM
Prof. V.M. MUROGOV   Institute of Physics and Power Engineering

SWITZERLAND
Mr. H. BAY   Nordostschweizerische Kraftwerke AG

UNITED KINGDOM
Mr. R. DODDS   British Nuclear Fuels plc
Mr. M. DUNN   British Nuclear Fuels plc
Dr. P. PARKES   British Nuclear Fuels plc
Mr. C. PROCTER   Nuclear Electric plc

UNITED STATES OF AMERICA
Dr. J.J. LAIDLER   Argonne National Laboratory
Dr. R.J. NEUHOLD   Department of Energy

IAEA
Dr. N. OI   Nuclear Fuel Cycle & Waste Management Division

EUROPEAN COMMISSION
Dr. A. DECRESSIN   Directorate-General XVII

OECD/NEA
Mr. G.H. STEVENS   Nuclear Development Division
Dr. N. ZARIMPAS (**Scientific Secretary**)   Nuclear Development Division

LIST OF ABBREVIATIONS, UNITS AND GLOSSARY OF TERMS

A

Ampere.

Actinide

A chemical element with atomic number between 89 (actinium) and 103 (lawrencium) located in the seventh period of the periodic table. Uranium and plutonium are other examples of actinide chemical elements.

AGR

British Advanced Gas-Cooled Reactor.

ATR

Japanese Advanced Thermal Reactor.

Back-end (of the fuel cycle)

Those nuclear fuel cycle processes and activities concerned with the treatment of spent fuel discharged from reactors including disposal of radioactive wastes.

BNFL

British Nuclear Fuels plc.

BOL

Beginning Of Life (*i.e.* beginning of the irradiation).

Bq

Becquerel.

Burn-up

The total energy released per unit mass of a nuclear fuel; it is commonly expressed in mega- or gigawatt-days per tonne (MWd/t or GWd/t).

BWR

Boiling-Water Reactor.

°C

Degrees Celsius.

CANDU

Canadian Deuterium-Uranium Reactor; a type of heavy-water reactor.

Ci

Curie.

Cladding

An external layer of material applied directly to nuclear fuel or other material that provides protection from a chemically reactive environment and containment of radioactive products produced during the irradiation of the composite. It may also provide structural support.

COGEMA

Compagnie Générale des Matières Nucléaires (France).

Conversion

The operation of altering the chemical form of a nuclear material to a form suitable for its end use.

Decommissioning

The work required for the planned permanent retirement of a plant from active service.

Direct disposal

Fuel cycle in which fuel goes through the reactor once; no spent fuel reprocessing is foreseen.

DOE

US Department of Energy.

efp

Equivalent full power.

EFR

European Fast Reactor.

Encapsulation

Processes associated with preparation of spent fuel for disposal.

Enrichment

  i) The fraction of atoms of a specified isotope in a mixture of isotopes of the same element when this fraction exceeds that in the naturally occurring mixture;

  ii) Any process by which the content of a specified isotope (uranium-235, etc.) in an element is increased.

EOL

End Of Life (*i.e.* end of the irradiation).

EURATOM

European Atomic Energy Community.

eV

Electron volt.

FA

Fuel Assembly.

Fabrication

The process of preparing nuclear fuel pellets, and cladding them to make fuel elements and the incorporation of elements into assemblies ready for the reactor.

FBR

Fast Breeder Reactor.

FGR

Fission Gas Release.

Fission

The physical process whereby the nucleus of a heavy atom is split into two (or, rarely, more) nuclei with masses of equal order of magnitude whose total mass is less than that of the original nucleus.

Fission products

Nuclides produced either by fission or by the subsequent radioactive decay of the nuclides thus formed.

FR

Fast Reactor.

Front-end (of the fuel cycle)

Those nuclear fuel cycle processes and activities concerned with the production of fuel for a reactor.

Fuel (nuclear)

Material containing fissile nuclides which, when placed in a reactor, enables a self-sustaining nuclear chain to be achieved.

Fuel cycle

The sequence of processing, manufacturing and transportation steps involved in producing fuel for a reactor, and in processing fuel discharged from the reactor including disposal of radioactive wastes.

g

Gram.

GWe

Gigawatt electric.

Half-life (radioactive)

For a single radioactive decay process, the time required for the activity to decrease to half its value by that process.

HLW

High-Level Waste.

HM

Heavy Metal (uranium, plutonium and other actinides in spent fuel).

HTGR

High Temperature Gas-Cooled Reactor.

HWR

Heavy-Water Reactor.

IAEA

International Atomic Energy Agency.

ILW

Intermediate-Level Waste.

Isotopes

Nuclides having the same atomic number (*i.e.* identical chemical element), but different mass numbers.

kg

Kilogram.

kWh

Kilowatt hour.

l

Litre.

lb

Pound.

LLW

Low-Level Waste.

LMFBR

Liquid Metal-Cooled Fast Breeder Reactor.

LMFR

Liquid Metal Fast Reactor.

Load factor

A ratio of the energy that is produced by a facility during the period considered to the energy that it could have produced at maximum capacity under continuous operation during the whole of that period.

LWR

Light-Water Reactor.

m

Meter.

MA

Minor Actinide.

MOX fuel

Mixed Oxide (uranium dioxide and plutonium dioxide) fuel.

MWd/t

Megawatt-day per tonne.

MWe

Megawatt electric.

MWt

Megawatt thermal.

NEA

OECD Nuclear Energy Agency.

NEA/NDC

Committee for Technical and Economic Studies on Nuclear Energy Development and the Fuel Cycle of the Nuclear Energy Agency.

OECD

Organisation for Economic Co-Operation and Development.

PNC

Power Reactor and Nuclear Fuel Development Corporation (Japan).

ppm

part per million (10^{-6}).

P&T

Partitioning and Transmutation.

Pu

Plutonium.

Puf

Plutonium fissile.

Pu(t)

All isotopes of plutonium, not only fissile.

PWR

Pressurised-Water Reactor.

R&D

Research and Development.

Reprocessing

A generic term for the chemical and mechanical processes applied to fuel elements discharged from a nuclear reactor. The purpose is to remove fission products and recover fissile (e.g. uranium-235, plutonium-239), fertile (e.g. uranium-238) and other valuable material.

SENA

Société d'Énergie Nucléaire Franco-Belge des Ardennes.

Spent fuel

Nuclear fuel removed from a reactor following irradiation.

t

Tonne.

THORP

Thermal Oxide Reprocessing Plant (UK).

TRU

Transuranic.

U

Uranium.

UOX

Uranium Oxide.

Waste management

All activities that are involved in the handling, treatment, conditioning, transportation, storage and disposal of waste.

VHLW

Vitrified High-Level Waste.

Waste repository

Prepared geological site suitable for permanent disposal of radioactive waste.

w/o

Weight per cent.

y

Year.

NUCLEAR POWER CAPACITIES IN OECD COUNTRIES[*]

Estimates of total and nuclear electricity capacity[d]

(Net GWe)

COUNTRY		1994 (Actual)			1995 (Actual)			2000		
		Total	Nuclear	%	Total	Nuclear	%	Total	Nuclear	%
Australia	(b)	37.3	0.0	0.0	37.6 (g)	0.0	0.0	40.1	0.0	0.0
Austria		17.0	0.0	0.0	17.1 (g)	0.0	0.0	18.1	0.0	0.0
Belgium		14.9	5.5	36.9	15.0	5.6	37.3	16.4	5.6	34.1
Canada		112.4	15.4	13.7	113.4	15.4	13.6	113.5	14.7	13.0
Denmark		9.0	0.0	0.0	9.0	0.0	0.0	9.6	0.0	0.0
Finland		13.1	2.3	17.6	13.5	2.3	17.0	16.5	2.6	15.8
France		107.0	58.5	54.7	107.5 (g)	58.5 (g)	54.4	113.3	64.3	56.8
Germany		114.3	22.7	19.9	117.5 (g)	22.7 (g)	19.3	118.0 (a)	23.1 (a)	19.6
Greece		8.9	0.0	0.0	8.9	0.0	0.0	11.8	0.0	0.0
Iceland		1.0	0.0	0.0	1.0	0.0	0.0	1.1	0.0	0.0
Ireland		3.9	0.0	0.0	4.0	0.0	0.0	4.7	0.0	0.0
Italy		64.1	0.0	0.0	65.8	0.0	0.0	75.7	0.0	0.0
Japan	(b,c,e)	195.1	38.4	19.7	200.1 (g)	39.3 (g)	19.6	234.5	42.9	18.3
Korea		28.8	7.6	26.4	32.1	8.6	26.8	52.7	13.7	26.0
Luxembourg		1.2	0.0	0.0	1.1	0.0	0.0	1.4	0.0	0.0
Mexico		31.7	0.7	2.1	33.0	1.3	4.0	37.3	1.3	3.5
Netherlands		18.7	0.5	2.7	19.2	0.5	2.6	27.9	0.5	1.8
New Zealand		7.7	0.0	0.0	7.7 (g)	0.0	0.0	8.4	0.0	0.0
Norway		27.5	0.0	0.0	27.6	0.0	0.0	27.8	0.0	0.0
Portugal		8.8	0.0	0.0	9.3 (g)	0.0	0.0	10.5	0.0	0.0
Spain		45.6	7.0	15.4	46.0 (g)	7.0 (g)	15.2	52.9	7.0	13.2
Sweden		34.5	10.0	29.0	34.1	10.0	29.3	34.8 (a)	10.0 (a)	28.7
Switzerland		15.7	3.1	19.5	15.8	3.1	19.4	16.3	3.2	19.6
Turkey		20.9	0.0	0.0	21.0	0.0	0.0	29.2	0.0	0.0
United Kingdom		68.9	11.7	17.0	69.8 (g)	12.9 (g)	18.5	82.4 (a)	12.9 (a)	15.7
United States		770.0	99.0	12.9	770.0	99.0	12.9	805.0	100.0	12.4
TOTAL		**1 777.9**	**282.3**	**15.9**	**1 797.0**	**286.2**	**15.9**	**1 959.9**	**301.8**	**15.4**
OECD America		914.1	115.1	12.6	916.4	115.7	12.6	955.8	116.0	12.1
OECD Europe		594.9	121.3	20.4	603.1	122.6	20.3	668.4	129.2	19.3
OECD Pacific	(f)	268.9	46.0	17.1	277.5	47.9	17.3	335.8	56.6	16.9

159

(continued)

(Net GWe)

COUNTRY		2005			2010			2015		
		Total	Nuclear	%	Total	Nuclear	%	Total	Nuclear	%
Australia	(b)	43.0	0.0	0.0	45.3	0.0	0.0			
Austria		18.5	0.0	0.0	19.0	0.0	0.0			
Belgium		17.6	5.6	31.8	17.8 (a)	5.6	31.5			
Canada		120.8	15.4	12.7	125.8	15.4	12.2	132.7	15.4	11.6
Denmark		8.5	0.0	0.0	N/A	0.0	0.0			
Finland		16.7	2.6	15.6	16.7	2.6	15.6	16.7	2.6	15.6
France		117.1	64.1	54.7	124.0	67.0	54.0			
Germany	(a)	121.0	23.1	19.1	123.4	23.1	18.7			
Greece		13.2	0.0	0.0	15.0	0.0	0.0			
Iceland		1.1	0.0	0.0	1.2	0.0	0.0	1.2	0.0	0.0
Ireland		5.5	0.0	0.0	6.4	0.0	0.0	7.5	0.0	0.0
Italy		79.8	0.0	0.0	86.1	0.0	0.0			
Japan	(b,c,e)	250.0 (a)	54.0 (a)	21.6	280.3	66.5	23.7			
Korea		67.9	18.7	27.5	79.5	26.3	33.1			
Luxembourg		1.4	0.0	0.0	1.4	0.0	0.0			
Mexico		44.5	1.3	2.9	56.8	1.3 (a)	2.3	72.5		
Netherlands		30.0	0.0 (a)	0.0	30.0 (a)	0.0 (a)	0.0			
New Zealand		8.9	0.0	0.0	9.4	0.0	0.0			
Norway		30.9	0.0	0.0	33.9	0.0	0.0			
Portugal		11.2	0.0	0.0	12.4	0.0	0.0			
Spain		57.5 (a)	8.5 (a)	14.8	68.2 (a)	10.0 (a)	14.7			
Sweden		34.8	10.0	28.7	30.5 (a)	0.0 (a)	0.0			
Switzerland		17.0	3.2	18.8	17.7	3.2	18.1	17.7 (g)	3.2 (g)	18.1
Turkey		42.3	1.0	2.4	60.1	2.0	3.3			
United Kingdom	(a)	90.9	9.8	10.8	100.1	7.4	7.4			
United States		834.0	100.0	12.0	877.0	91.0	10.4	935.0	61.0	6.5
TOTAL		**2 084.1**	**317.3**	**15.2**	**2 238.0**	**321.4**	**14.4**			
OECD America		999.3	116.7	11.7	1 059.6	107.7	10.2			
OECD Europe		715.0	127.9	17.9	763.9	120.9	15.8			
OECD Pacific	(f)	369.8	72.7	19.7	414.5	92.8	22.4			

(*)   Table extracted from the 1996 NEA Nuclear Energy Data, the *"Brown Book",* OECD/NEA, 1996.

(a)   Secretariat estimate.
(b)   For fiscal year (July-June for Australia, April-March for Japan).
(c)   Gross data converted to net by Secretariat.
(d)   Including electricity generated by the user (autoproduction), unless stated otherwise.
(e)   Excluding electricity generated by the user (autoproduction).
(f)   Including data for Korea.
(g)   Provisional data.
N/A   Not Available.

MAIN SALES OUTLETS OF OECD PUBLICATIONS
PRINCIPAUX POINTS DE VENTE DES PUBLICATIONS DE L'OCDE

AUSTRALIA – AUSTRALIE
D.A. Information Services
648 Whitehorse Road, P.O.B 163
Mitcham, Victoria 3132   Tel. (03) 9210.7777
                          Fax: (03) 9210.7788

AUSTRIA – AUTRICHE
Gerold & Co.
Graben 31
Wien 1   Tel. (0222) 533.50.14
                          Fax: (0222) 512.47.31.29

BELGIUM – BELGIQUE
Jean De Lannoy
Avenue du Roi, Koningslaan 202
B-1060 Bruxelles   Tel. (02) 538.51.69/538.08.41
                          Fax: (02) 538.08.41

CANADA
Renouf Publishing Company Ltd.
5369 Canotek Road
Unit 1
Ottawa, Ont. K1J 9J3   Tel. (613) 745.2665
                          Fax: (613) 745.7660

Stores:
71 1/2 Sparks Street
Ottawa, Ont. K1P 5R1   Tel. (613) 238.8985
                          Fax: (613) 238.6041

12 Adelaide Street West
Toronto, QN M5H 1L6   Tel. (416) 363.3171
                          Fax: (416) 363.5963

Les Éditions La Liberté Inc.
3020 Chemin Sainte-Foy
Sainte-Foy, PQ G1X 3V6   Tel. (418) 658.3763
                          Fax: (418) 658.3763

Federal Publications Inc.
165 University Avenue, Suite 701
Toronto, ON M5H 3B8   Tel. (416) 860.1611
                          Fax: (416) 860.1608

Les Publications Fédérales
1185 Université
Montréal, QC H3B 3A7   Tel. (514) 954.1633
                          Fax: (514) 954.1635

CHINA – CHINE
Book Dept., China National Publications
Import and Export Corporation (CNPIEC)
16 Gongti E. Road, Chaoyang District
Beijing 100020   Tel. (10) 6506-6688 Ext. 8402
                          (10) 6506-3101

CHINESE TAIPEI – TAIPEI CHINOIS
Good Faith Worldwide Int'l. Co. Ltd.
9th Floor, No. 118, Sec. 2
Chung Hsiao E. Road
Taipei   Tel. (02) 391.7396/391.7397
                          Fax: (02) 394.9176

**CZECH REPUBLIC –
RÉPUBLIQUE TCHÈQUE**
National Information Centre
NIS – prodejna
Konviktská 5
Praha 1 – 113 57   Tel. (02) 24.23.09.07
                          Fax: (02) 24.22.94.33
E-mail: nkposp@dec.niz.cz
Internet: http://www.nis.cz

DENMARK – DANEMARK
Munksgaard Book and Subscription Service
35, Nørre Søgade, P.O. Box 2148
DK-1016 København K   Tel. (33) 12.85.70
                          Fax: (33) 12.93.87

J. H. Schultz Information A/S,
Herstedvang 12,
DK – 2620 Albertslung   Tel. 43 63 23 00
                          Fax: 43 63 19 69
Internet: s-info@inet.uni-c.dk

EGYPT – ÉGYPTE
The Middle East Observer
41 Sherif Street
Cairo   Tel. (2) 392.6919
                          Fax: (2) 360.6804

FINLAND – FINLANDE
Akateeminen Kirjakauppa
Keskuskatu 1, P.O. Box 128
00100 Helsinki

Subscription Services/Agence d'abonnements :
P.O. Box 23
00100 Helsinki   Tel. (358) 9.121.4403
                          Fax: (358) 9.121.4450

*FRANCE
OECD/OCDE
Mail Orders/Commandes par correspondance :
2, rue André-Pascal
75775 Paris Cedex 16   Tel. 33 (0)1.45.24.82.00
                        Fax: 33 (0)1.49.10.42.76
                        Telex: 640048 OCDE
Internet: Compte.PUBSINQ@oecd.org

Orders via Minitel, France only/
Commandes par Minitel, France exclusivement :
36 15 OCDE

OECD Bookshop/Librairie de l'OCDE :
33, rue Octave-Feuillet
75016 Paris   Tel. 33 (0)1.45.24.81.81
                         33 (0)1.45.24.81.67
Dawson
B.P. 40
91121 Palaiseau Cedex   Tel. 01.89.10.47.00
                          Fax: 01.64.54.83.26

Documentation Française
29, quai Voltaire
75007 Paris   Tel. 01.40.15.70.00

Economica
49, rue Héricart
75015 Paris   Tel. 01.45.78.12.92
                          Fax: 01.45.75.05.67

Gibert Jeune (Droit-Économie)
6, place Saint-Michel
75006 Paris   Tel. 01.43.25.91.19

Librairie du Commerce International
10, avenue d'Iéna
75016 Paris   Tel. 01.40.73.34.60

Librairie Dunod
Université Paris-Dauphine
Place du Maréchal-de-Lattre-de-Tassigny
75016 Paris   Tel. 01.44.05.40.13

Librairie Lavoisier
11, rue Lavoisier
75008 Paris   Tel. 01.42.65.39.95

Librairie des Sciences Politiques
30, rue Saint-Guillaume
75007 Paris   Tel. 01.45.48.36.02

P.U.F.
49, boulevard Saint-Michel
75005 Paris   Tel. 01.43.25.83.40

Librairie de l'Université
12a, rue Nazareth
13100 Aix-en-Provence   Tel. 04.42.26.18.08

Documentation Française
165, rue Garibaldi
69003 Lyon   Tel. 04.78.63.32.23

Librairie Decitre
29, place Bellecour
69002 Lyon   Tel. 04.72.40.54.54

Librairie Sauramps
Le Triangle
34967 Montpellier Cedex 2 Tel. 04.67.58.85.15
                           Fax: 04.67.58.27.36

A la Sorbonne Actual
23, rue de l'Hôtel-des-Postes
06000 Nice   Tel. 04.93.13.77.75
                          Fax: 04.93.80.75.69

GERMANY – ALLEMAGNE
OECD Bonn Centre
August-Bebel-Allee 6
D-53175 Bonn   Tel. (0228) 959.120
                      Fax: (0228) 959.12.17

GREECE – GRÈCE
Librairie Kauffmann
Stadiou 28
10564 Athens   Tel. (01) 32.55.321
                          Fax: (01) 32.30.320

HONG-KONG
Swindon Book Co. Ltd.
Astoria Bldg. 3F
34 Ashley Road, Tsimshatsui
Kowloon, Hong Kong   Tel. 2376.2062
                          Fax: 2376.0685

HUNGARY – HONGRIE
Euro Info Service
Margitsziget, Európa Ház
1138 Budapest   Tel. (1) 111.60.61
                          Fax: (1) 302.50.35
E-mail: euroinfo@mail.matav.hu
Internet: http://www.euroinfo.hu//index.html

ICELAND – ISLANDE
Mál og Menning
Laugavegi 18, Pósthólf 392
121 Reykjavik   Tel. (1) 552.4240
                          Fax: (1) 562.3523

INDIA – INDE
Oxford Book and Stationery Co.
Scindia House
New Delhi 110001   Tel. (11) 331.5896/5308
                     Fax: (11) 332.2639
E-mail: oxford.publ@axcess.net.in

17 Park Street
Calcutta 700016   Tel. 240832

INDONESIA – INDONÉSIE
Pdii-Lipi
P.O. Box 4298
Jakarta 12042   Tel. (21) 573.34.67
                          Fax: (21) 573.34.67

IRELAND – IRLANDE
Government Supplies Agency
Publications Section
4/5 Harcourt Road
Dublin 2   Tel. 661.31.11
                          Fax: 475.27.60

ISRAEL – ISRAËL
Praedicta
5 Shatner Street
P.O. Box 34030
Jerusalem 91430   Tel. (2) 652.84.90/1/2
                          Fax: (2) 652.84.93

R.O.Y. International
P.O. Box 13056
Tel Aviv 61130   Tel. (3) 546 1423
                          Fax: (3) 546 1442
E-mail: royil@netvision.net.il

Palestinian Authority/Middle East:
INDEX Information Services
P.O.B. 19502
Jerusalem   Tel. (2) 627.16.34
                          Fax: (2) 627.12.19

ITALY – ITALIE
Libreria Commissionaria Sansoni
Via Duca di Calabria, 1/1
50125 Firenze   Tel. (055) 64.54.15
                          Fax: (055) 64.12.57
E-mail: licosa@ftbcc.it

Via Bartolini 29
20155 Milano   Tel. (02) 36.50.83

Editrice e Libreria Herder
Piazza Montecitorio 120
00186 Roma   Tel. 679.46.28
                          Fax: 678.47.51

Libreria Hoepli
Via Hoepli 5
20121 Milano   Tel. (02) 86.54.46
                          Fax: (02) 805.28.86

Libreria Scientifica
Dott. Lucio de Biasio 'Aeiou'
Via Coronelli, 6
20146 Milano   Tel. (02) 48.95.45.52
    Fax: (02) 48.95.45.48

JAPAN – JAPON
OECD Tokyo Centre
Landic Akasaka Building
2-3-4 Akasaka, Minato-ku
Tokyo 107   Tel. (81.3) 3586.2016
    Fax: (81.3) 3584.7929

KOREA – CORÉE
Kyobo Book Centre Co. Ltd.
P.O. Box 1658, Kwang Hwa Moon
Seoul   Tel. 730.78.91
    Fax: 735.00.30

MALAYSIA – MALAISIE
University of Malaya Bookshop
University of Malaya
P.O. Box 1127, Jalan Pantai Baru
59700 Kuala Lumpur
Malaysia   Tel. 756.5000/756.5425
    Fax: 756.3246

MEXICO – MEXIQUE
OECD Mexico Centre
Edificio INFOTEC
Av. San Fernando no. 37
Col. Toriello Guerra
Tlalpan C.P. 14050
Mexico D.F.   Tel. (525) 528.10.38
    Fax: (525) 606.13.07
E-mail: ocde@rtn.net.mx

NETHERLANDS – PAYS-BAS
SDU Uitgeverij Plantijnstraat
Externe Fondsen
Postbus 20014
2500 EA's-Gravenhage   Tel. (070) 37.89.880
Voor bestellingen: Fax: (070) 34.75.778

Subscription Agency/   Agence d'abonnements :
SWETS & ZEITLINGER BV
Heereweg 347B
P.O. Box 830
2160 SZ Lisse   Tel. 252.435.111
    Fax: 252.415.888

**NEW ZEALAND –
NOUVELLE-ZÉLANDE**
GPLegislation Services
P.O. Box 12418
Thorndon, Wellington   Tel. (04) 496.5655
    Fax: (04) 496.5698

NORWAY – NORVÈGE
NIC INFO A/S
Ostensjoveien 18
P.O. Box 6512 Etterstad
0606 Oslo   Tel. (22) 97.45.00
    Fax: (22) 97.45.45

PAKISTAN
Mirza Book Agency
65 Shahrah Quaid-E-Azam
Lahore 54000   Tel. (42) 735.36.01
    Fax: (42) 576.37.14

PHILIPPINE – PHILIPPINES
International Booksource Center Inc.
Rm 179/920 Cityland 10 Condo Tower 2
HV dela Costa Ext cor Valero St.
Makati Metro Manila   Tel. (632) 817 9676
    Fax: (632) 817 1741

POLAND – POLOGNE
Ars Polona
00-950 Warszawa
Krakowskie Prezdmiescie 7   Tel. (22) 264760
    Fax: (22) 265334

PORTUGAL
Livraria Portugal
Rua do Carmo 70-74
Apart. 2681
1200 Lisboa   Tel. (01) 347.49.82/5
    Fax: (01) 347.02.64

SINGAPORE – SINGAPOUR
Ashgate Publishing
Asia Pacific Pte. Ltd
Golden Wheel Building, 04-03
41, Kallang Pudding Road
Singapore 349316   Tel. 741.5166
    Fax: 742.9356

SPAIN – ESPAGNE
Mundi-Prensa Libros S.A.
Castelló 37, Apartado 1223
Madrid 28001   Tel. (91) 431.33.99
    Fax: (91) 575.39.98
E-mail: mundiprensa@tsai.es
Internet: http://www.mundiprensa.es

Mundi-Prensa Barcelona
Consell de Cent No. 391
08009 – Barcelona   Tel. (93) 488.34.92
    Fax: (93) 487.76.59

Libreria de la Generalitat
Palau Moja
Rambla dels Estudis, 118
08002 – Barcelona
    (Suscripciones) Tel. (93) 318.80.12
    (Publicaciones) Tel. (93) 302.67.23
    Fax: (93) 412.18.54

SRI LANKA
Centre for Policy Research
c/o Colombo Agencies Ltd.
No. 300-304, Galle Road
Colombo 3   Tel. (1) 574240, 573551-2
    Fax: (1) 575394, 510711

SWEDEN – SUÈDE
CE Fritzes AB
S-106 47 Stockholm   Tel. (08) 690.90.90
    Fax: (08) 20.50.21

For electronic publications only/
Publications électroniques seulement
STATISTICS SWEDEN
Informationsservice
S-115 81 Stockholm   Tel. 8 783 5066
    Fax: 8 783 4045

Subscription Agency/Agence d'abonnements :
Wennergren-Williams Info AB
P.O. Box 1305
171 25 Solna   Tel. (08) 705.97.50
    Fax: (08) 27.00.71

Liber distribution
Internatinal organizations
Fagerstagatan 21
S-163 52 Spanga

SWITZERLAND – SUISSE
Maditec S.A. (Books and Periodicals/Livres
et périodiques)
Chemin des Palettes 4
Case postale 266
1020 Renens VD 1   Tel. (021) 635.08.65
    Fax: (021) 635.07.80

Librairie Payot S.A.
4, place Pépinet
CP 3212
1002 Lausanne   Tel. (021) 320.25.11
    Fax: (021) 320.25.14

Librairie Unilivres
6, rue de Candolle
1205 Genève   Tel. (022) 320.26.23
    Fax: (022) 329.73.18

Subscription Agency/Agence d'abonnements :
Dynapresse Marketing S.A.
38, avenue Vibert
1227 Carouge   Tel. (022) 308.08.70
    Fax: (022) 308.07.99

See also – Voir aussi :
OECD Bonn Centre
August-Bebel-Allee 6
D-53175 Bonn (Germany)   Tel. (0228) 959.120
    Fax: (0228) 959.12.17

THAILAND – THAÏLANDE
Suksit Siam Co. Ltd.
113, 115 Fuang Nakhon Rd.
Opp. Wat Rajbopith
Bangkok 10200   Tel. (662) 225.9531/2
    Fax: (662) 222.5188

**TRINIDAD & TOBAGO, CARIBBEAN
TRINITÉ-ET-TOBAGO, CARAÏBES**
Systematics Studies Limited
9 Watts Street
Curepe
Trinidad & Tobago, W.I.   Tel. (1809) 645.3475
    Fax: (1809) 662.5654
E-mail: tobe@trinidad.net

TUNISIA – TUNISIE
Grande Librairie Spécialisée
Fendri Ali
Avenue Haffouz Imm El-Intilaka
Bloc B 1 Sfax 3000   Tel. (216-4) 296 855
    Fax: (216-4) 298.270

TURKEY – TURQUIE
Kültür Yayinlari Is-Türk Ltd.
Atatürk Bulvari No. 191/Kat 13
06684 Kavaklidere/Ankara
    Tel. (312) 428.11.40 Ext. 2458
    Fax : (312) 417.24.90
Dolmabahce Cad. No. 29
Besiktas/Istanbul   Tel. (212) 260 7188

UNITED KINGDOM – ROYAUME-UNI
The Stationery Office Ltd.
Postal orders only:
P.O. Box 276, London SW8 5DT
Gen. enquiries   Tel. (171) 873 0011
    Fax: (171) 873 8463

The Stationery Office Ltd.
Postal orders only:
49 High Holborn, London WC1V 6HB
Branches at: Belfast, Birmingham, Bristol,
Edinburgh, Manchester

UNITED STATES – ÉTATS-UNIS
OECD Washington Center
2001 L Street N.W., Suite 650
Washington, D.C. 20036-4922 Tel. (202) 785.6323
    Fax: (202) 785.0350
Internet: washcont@oecd.org

Subscriptions to OECD periodicals may also be
placed through main subscription agencies.

Les abonnements aux publications périodiques de
l'OCDE peuvent être souscrits auprès des
principales agences d'abonnement.

Orders and inquiries from countries where Distribu-
tors have not yet been appointed should be sent to:
OECD Publications, 2, rue André-Pascal, 75775
Paris Cedex 16, France.

Les commandes provenant de pays où l'OCDE n'a
pas encore désigné de distributeur peuvent être
adressées aux Éditions de l'OCDE, 2, rue André-
Pascal, 75775 Paris Cedex 16, France.

12-1996

OECD PUBLICATIONS, 2, rue André-Pascal, 75775 PARIS CEDEX 16
PRINTED IN FRANCE
(66 97 01 1P) ISBN 92-64-15410-8 – No. 49221 1997